# The Top 10 Myths about Evolution

# The Top 10 Myths about Evolution

Cameron M. Smith
and Charles Sullivan

Illustrations by Gerald Fried

59 John Glenn Drive
Amherst, New York 14228-2197

Published 2007 by Prometheus Books



Inquiries should be addressed to
Prometheus Books
59 John Glenn Drive
Amherst, New York 14228–2197
VOICE: 716–691–0133, ext. 207
FAX: 716–564–2711
WWW.PROMETHEUSBOOKS.COM

11 10 09 08 07 5 4 3 2

Library of Congress Cataloging-in-Publication Data

Smith, Cameron McPherson, 1967–
The top ten myths about evolution / by Cameron McPherson Smith and Charles Sullivan.
p. cm.
Includes bibliographical references and index.
ISBN-13: 978–1–59102–479–8 (pbk. : alk. paper)
ISBN-10: 1–59102–479–X (pbk. : alk. paper)
1. Evolution (Biology) I. Sullivan, Charles, 1963– II. Title.

QH366.2.S54 2006
576.8—dc22

2006022768

Printed in the United States of America on acid-free paper

# Contents

MUS
VERITAS • HONORQUE

# Acknowledgments

CAMERON M. SMITH

I thank my parents, Professors Donald E. and Margit J. Smith, for their support of my education, worldwide and over twenty years. I would also like to thank Professor Kenneth M. Ames for teaching me, early on, that I should be most skeptical of my most favorite theories. For their fascinating and stimulating conversation I thank, in no particular order, Dr. Farid Rahemtulla, Dr. Evan T. Davies, Mark J. and D. Julian Smith, Rudy Reimer, Dr. Ramona Maraj, W. McRee Anderson, Peter Panek, Professor Brian Hayden, Charles Sullivan, Professors Alan Bilsborough, Anthony Harding, Marc Feldesman, Virginia Butler, and Philip Tobias, Todd Olson, John Mitchell, Ms. Christine Calfas, Dr. Yin Lam, John and Annie Haslett, and Dr. John Lonsbury. Finally, I thank Charles Sullivan for his good cheer during our sometimes "challenging" bouts of coediting.

CHARLES SULLIVAN

I would first like to thank my mother, Dr. Pat Madden Sullivan, for introducing me to the importance of science and learning in my youth, and for helping to pique my curiosity about nature through providing children's science books, frequent visits to the dinosaur bones at the natural history museum, and a vision of the grandeur of the universe at the local planetarium. I would also like to thank Professors Larry Bowlden and Tony Wolk for encouraging my intellectual development and my love of lifelong learning. Other professors who played no small part are Byron Haines, Michael Philips, and Carol Franks. I also thank Nora, Dimitris, and Nicholas Houndalas for the family love they are creating. And finally, I thank my old friend Cameron McPherson Smith for his truly untiring inspiration and camaraderie.

We would both like to thank Ann Druyan for her encouraging correspondence as well as her promotional comment for this book, and Richard Dawkins for his (mostly favorable) review of the original *Skeptical Inquirer* article on which it is based, and for pointing out an important oversight. We also thank the fine staff at Prometheus Books for taking on our vision (Steven L. Mitchell), creating the art to convey it visually (Gerald Fried), and helping to edit the words (Joe Gramlich). We could not have written this book without the hard work of generations of scientists and philosophers. Any mistakes are our responsibility.

# Introduction

Since humanity first became self-aware, every culture has tried to account for the diversity of life on Earth, conjuring a dizzying catalog of explanations for the natural world.[1] Babylonians, Byzantines, Sumerians, and Samoans alike devised ingenious myths, all with one thing in common. In every case, plants and animals were created by supernatural beings—gods—in order to fulfill some divine purpose. For the Babylonians, humanity itself was created as a slave race, a servant to the supreme deity Marduk. These mythological accounts of creation molded the ancient world, and people lived and died for them.

Twenty-five hundred years ago, however, a new approach was being tried out in Greece,[2] an approach that favored natural rather than supernatural explanations. This worked for physics, chemistry, and other sciences, but it was only fully realized when it was applied to the world of living things. That happened when Darwin published *On the Origin of Species* in 1859, transforming the Western world's most basic concepts about the origins and nature of species.

Darwin's ideas were simple, powerful, and they explained so much that today evolution is the foundation of the life sciences. But after nearly 150 years, evolution is still commonly misunderstood by the general public, particularly in the United States, where a number of myths about evolution arisen.[3] Since evolution is central to biology, and biology is the best explanation of living systems, evolution itself is a guide to the decisions we make about critical issues, such as protection of endangered species, stem-cell research, and genetic engineering. Because evolution touches everyone, through biology, it's important to get it right.

But many don't get it right because of the power of these myths. Many of these myths are based on ignorance, for which the best remedy is knowledge. Some are perpetuated because evolution is inadequately taught in high schools, although many teachers do an excellent job. The media also contribute a fair share of confusion, from the misleading depictions of evolution sometimes found in science fiction movies to the poverty of science programming on commercial television. There are even more troubling forces at work in America that make it hard to debunk these myths, such as the people who believe that their religious texts and traditions can provide us with scientific answers even though those answers come from prescientific times.

All of these influences lead to a distorted picture of how evolution works. And this is revealed in surveys showing that only about half of Americans realize that humans and dinosaurs never lived side by side; and about the same number reject the idea that humans developed from earlier species of animals.[4]

In this book, we identify the top ten myths about evolution, and we show why they're wrong. In 2001 the ninety-seven-year-old biologist Ernst Mayr published a wonderful introduction to evolution, *What Evolution Is*.[5] In a way, *The Top 10 Myths about Evolution* explains the opposite: What evolution *isn't*. Each myth-busting chapter exposes the flaws behind a myth and provides examples that can help readers refute these myths when they encounter them.

"Myth One: Survival of the Fittest" shows how common descriptions of nature as a bloody world of competition have distorted our understanding of what fitness really means in evolution.

"Myth Two: It's Just a Theory" shows what scientific theories really are, describes their power to explain observations, and explains why calling evolution "just a theory" is a mistake. This chapter also explains in plain language how evolution works.

"Myth Three: The Ladder of Progress" explores the myth that humans are the goal of evolution, that there's an inevitable progression from lower life-forms leading to human beings.

"Myth Four: The Missing Link" shows how the idea of a single missing link between species distorts the reality of the changing, fluid nature of evolution.

"Myth Five: Evolution Is Random" clarifies what randomness means and shows how evolution is not random, yet has the appearance of design without being designed.

"Myth Six: People Come from Monkeys" clarifies our true relationship to monkeys and to chimpanzees—our closest living nonhuman cousins—and explains how we evolved into our modern form.

"Myth Seven: Nature's Perfect Balance" explores the misunderstandings of exactly what ecological balance is, and what role—if any—evolution plays in maintaining environmental equilibrium.

"Myth Eight: Creationism Disproves Evolution" shows how creationism's argument for a young Earth, and its challenge to the scientific evidence supporting evolution, is flawed.

"Myth Nine: Intelligent Design Is Science" explains why intelligent design—creationism's dressed-up twin—ultimately fails to live up to the standards of genuine science, and exposes the political agenda driving this movement.

"Myth Ten: Evolution Is Immoral" explores the moral implications of evolution, showing how the theory has been distorted to support the ideology that *might makes right*, as well as exploring how many of our moral sentiments may have evolutionary roots.

## Notes

1. Although archaeologists aren't certain when humanity became self-aware, the origins of myth, ritual, and symbol are clearly traced to more than thirty thousand years ago. See S. Mithen, *The Prehistory of the Mind: A Search for the Origins of Art, Religion and Science* (London: Thames & Hudson, 1996).

2. Although nearly a century old, W. Durant's *The Story of Philosophy* (1926) is still a fresh discussion of the early Greeks. Their approach, based in seeking natural rather than supernatural explanations, is the foundation of science today, and it applies to everyone. Physics and engineering keep planes in the air, not Aeolus, the Greek god of wind. Since physics and biology build from the same base, if you believe in the physics that keeps planes in the air, you must also believe in the system of knowledge that explains evolution. You can't have "a little science": it's all or nothing. See W. Durant, *The Story of Philosophy* (New York: Simon & Schuster, 1926).

3. See C. Sullivan and C. M. Smith, "Getting the Monkey off Darwin's Back: Four Common Myths about Evolution," *Skeptical Inquirer* (May/June 2005): 43–48.

4. National Science Foundation, Division of Science Resources and Statistics, "Science and Technology: Public Attitudes and Understanding," in *Science and Engineering Indicators*, chap. 7, pp. 15–16 (2004), http://www.nsf.gov/statistics/seind 04/pdf/c07.pdf (accessed May 5, 2005).

5. E. Mayr, *What Evolution Is* (New York: Basic Books, 2001).

## Myth One

# Survival of the Fittest

"Survival of the fittest" is the most commonly used phrase drafted into everyday speech from the theory of evolution. Flipping through television channels, we see a lion bearing down on a gazelle, a boxer pummeling his opponent, bighorn sheep clashing horns: we nod and smugly think, "There, see? *Survival of the fittest*; the order of nature." And it seems clear enough: for all we can tell, the strong *do* survive. It would be crazy to think otherwise, considering what we've learned about the natural world from mass media.

But mass media, of course, is about drama and unfolding stories, and every dramatist knows that without conflict, you have no story. And so the natural world has been dressed up as a vast and violent landscape of competition—the ultimate reality show, where real blood can be shed. Can the antelope corner tighter than the lion? Can the species survive? Will the "balance of nature" be upset? Television has taken Tennyson to heart, portraying nature as "red in tooth and claw," a world of savage predation, where survival of the fittest is the primary law.[1]

This is all very exciting, but it's a vision of the natural world focused almost entirely on dramatic competition. If we shift our focus, though, it's clear that there's an entire world of plant and animal relationships that aren't dominated by violent competition. For example, aside from the obvious eating of prey by predators, most animals generally leave each other alone, particularly if they're after different kinds of food. And of course there's symbiosis, in which species interactions are mutually beneficial. Nevertheless, on tonight's Animal Planet schedule, we have plenty of shows on big, scary, ferocious animals (polar bears, lions, and spitting cobras), while only one herbivore is getting airtime—the poor old wildebeest. And what about plants? Their lack of bloody teeth and claws might account for the lack of a "Plant Planet" channel.[2]

Clearly, the popular focus on competition has led to a portrayal of nature as a metaphorical battlefield, where all that matters is your ability to wage war, outstrip your opponent, and beat down your immediate peers: the "survival of the fittest."[3] But, as usual, nature is far too complex to be reduced to this absolute, bumper-sticker slogan. Let's see if we can clear up the confusion caused by the survival of the fittest myth, starting with Darwin himself.

## WHAT DID DARWIN MEAN BY "SURVIVAL OF THE FITTEST"?

We know how popular culture portrays survival of the fittest, but what did Darwin himself have to say about it? In 1872 he wrote: "[The] preservation of favourable individual differences and variations, and the destruction of those which are injurious, I have called Natural Selection, or the Survival of the Fittest."[4]

So, according to Darwin himself, "survival of the fittest" and "natural selection" are basically the same thing—both phrases tell us that in any population, those individuals with characteristics well suited to their environment tend to be preserved, while those less well suited tend to die off. What's most important is that Darwin doesn't specify anything here about the characteristics we often casually associate with the idea that only the strong survive, such as brute force. In fact, Darwin doesn't mention *any specific characteristics* (like bulging muscles, sharp teeth, or a keen sense for smelling blood) equating to fitness. That should raise the first red flag when people speak casually about survival of the fittest, because we can immediately ask, *fittest in what sense* or *fittest in what environment*? Clearly, no single ability or physical characteristic makes any life-form "fit" in every environment.

For example, the long-finned eel is expert at snapping up dragonfly larvae, snails, and small fish that live in New Zealand's lakes, and it's in these lakes where they grow fat (up to eighty pounds) and happy, with few predators. They're kings of their cool, clear-water environment. But before they can grow fat and happy, they have to survive a perilous migration from their birthplace in the Pacific Ocean, and that often requires them to squirm across dry land; most make it, but dry spells occasionally strand armies of them in sticky mudflats. Slightly less rainfall than average can make carcasses of these would-be kings.

Humans, it's often thought, are immune to the rigors of nature—the "selective pressures" that any life-form must endure to survive—because we've mastered living in so many environments. But newborns can't walk or crawl, have no teeth or claws, and can only really eat breast milk, so actually there are many selective pressures that can quickly kill our offspring. At birth, in fact, we're among the most helpless of animals, entirely dependent on our parents. Our specialty as a species isn't *brawn* but *brain*.

Obviously, then, we need to refine our concept of the survival of the fittest. It's pretty clear what "survival" refers to. For the individual, it means staying alive. For a species, it means that enough individuals stay alive long enough to have offspring, and perhaps long enough to care for those offspring. The problem here is the word *fittest*. Fittest, of course, means the most fit, so let's examine the concept of *fitness* itself.

## Fitness

What is fitness? From the perspective of population genetics, fitness is basically the statistical likelihood that you'll have offspring, a cosmic wager on your genetic prospects as an individual.[5] Fitness isn't any single physical characteristic of an organism, like musculature or tooth size; it's a *measure of an individual's reproductive potential*, whether that individual is bat, buffalo, or bamboo.

While population geneticists calculate theoretical fitness for lab studies, calculating that exact probability for any individual in the real world would take a bank of supercomputers going full-tilt, 24/7. Why? Because the probability that you'll have offspring can be affected by so many factors—each called a *selective pressure*—that your fitness score is constantly changing.[6] An individual's fitness fluctuates so much that no bookie would dare wager on it, and that's because no bookie, or supercomputer, could ever estimate the range and the effects of selective agents that modify your fitness score from one moment to the next.

For example, imagine that you're a male Neanderthal living with your family group about one hundred thousand years ago.[7] You inhabit a cave overlooking a broad mountain valley in a place that will one day be called France. What are the selective agents that affect your fitness, your likelihood of having offspring? Brute force is an asset (you don't know it, but you have twice the strength of a modern human), but is brute strength really enough to keep you alive? You have a lot of concerns. You have to find water and food, and since storage hasn't really been invented yet, every day you hike down from your cave, embarking on the food quest. You know that around the time certain plants begin to bloom, herds of reindeer usually cross the river that runs down your valley; perhaps today it's time to lay in wait for them. But sometimes they don't come at the right time (maybe they've been intercepted by wolves, or diverted by other Neanderthals), and that can be disastrous; your fitness score might plummet. There's also the danger that *you'll* become lunch; cave lions, five-hundred-pound predators bigger than modern lions, are out there, and they know what you do—they're smart, they remember. And of course there's the danger of running into other Neanderthals, who might not like your intruding into their hunting ground. There's also the fire to worry about; it's been raining for a week, and if the fire dies, you're going to have a hard time restarting it. And you need to find a mate, but at twenty-eight, you're already pretty old. Your fitness, in fact, decreases daily now, as the likelihood of your death increases simply as a function of time. And what about the unknowns in your world? What if you get sick for a reason you'll never fathom? And could you predict that one day a new and different variety of people (modern humans) is going to appear at the foot of your valley with better tools and smarter brains? Sometimes, selective pressures leap at you from the abyss of ignorance.

It's clear that at any given time the selective pressures impinging on Neanderthals—or any species—are numerous, complexly interconnected, and largely unpredictable. Objectively, it's easy to *define* fitness as "an organism's likelihood of reproduction," but our Neanderthal example has shown us that this likelihood is a probability figure—perhaps with an infinitely long decimal—that *changes from moment to moment*. Fitness is fluid, not concrete.

Imagine a Cosmic Computer tracking your fitness score, a single probability figure, from 0–100 percent on a little red display that flutters and flashes from moment to moment as selective pressures increase or decrease your chances of passing your genes on to the next generation of your offspring. As the properties of your selective environment shift and slide, riding seasonal tides or being jarred by catastrophes, your fitness score, blinking away on the Cosmic Computer, shows their effects. Sometimes you're favored by a change in your environment (as when our Neanderthal comes across a dying mammoth), other times you're not (as when our Neanderthal is driven away from a kill by a pack of hyenas). No single characteristic, like brute strength or agility or even intelligence, is a savior, because the pressures on you can differ from moment to moment, from place to place, or even from one generation to the next.

Clearly, then, fitness is relative to any life-form's environment at large, and this leads us to the term *selective environment*. Let's see if examining the selective environment can help us sharpen focus on the fuzzy phrase "survival of the fittest."

## Selective Environments

Another way to envision fitness is to think of it as a measure of the *closeness of fit* between you (and "you" may be a sunflower, a slime mold, a rabbit, or a wren) and your selective environment. Among your seedy, slimy, furry, or feathered counterparts, everyone has a slightly different fit to their environment, because it's very rare that individuals of any population are exact clones; there is almost always some variation, however slight, between individuals.[8] Clearly, the fittest of your population is the one *best suited* to that environment (and therefore more likely to reproduce), and the one with a poor fit—for example, a fly born without wings—is less fit, and less likely to have offspring. This way of envisioning fitness is only a little less nebulous than the Cosmic Computer's ever-changing readout, but it's somehow a little more concrete. You can *see it*, for example, in the poor, wingless fly, as you can in many cases in the natural world.[9]

Consider cheetahs. They appear to have had a catastrophic population crash around ten thousand years ago, after which extensive inbreeding of the surviving population created a number of problems, including abnormalities in their sperm, high infant mortality, and susceptibility to a variety of diseases.[10] Most visibly, some cheetahs are born with slightly kinked or curled tails, which is a big problem because normally, the cheetah's tail is used as a sort of rudder, or counterbalance, during its seventy-mile-per-hour sprints in pursuit of prey. Cheetahs born with an abnormal tail simply won't have as much control, or hunting success, as those with a sleek, controllable tail. The difference in fitness here, in *fit between individual cheetahs and their selective environment*, is obvious, and it has nothing to do with one-on-one combat between cheetahs, or with their ability to slay each other in some television producer's gory arena of competition. Here, an important factor tinkering with the fitness score is simply whether a cheetah is born with a kinked or straight tail.

We can see selection at work in even less outwardly competitive realms. Let's say you're a woodpecker, equipped with a stout beak and a compact skull that firmly cups your brain, preventing a chronic migraine. But you're a little hungry because, sometimes, when you listen for grubs after jackhammering a tree with your beak, you don't hear anything. Off you fly, hungry and disappointed; it's only happened twice today, but that's enough to be frustrating. You have no idea that this is a simple hearing defect, and that your neighbor has better hearing; all you know is you see him getting fat on the grubs you can't find. The blunt facts are that, simply because of a hearing defect, your chances of finding grubs are reduced, and that your health may suffer and your chances of finding a mate—for instance, by hammering long and hard at a tree to attract one—is lower than your neighbor's. In this case, an apparently minute factor of the environment—the level of noise produced by burrowing under-bark grubs—is an important aspect of the woodpecker's selective environment. The subtlety of this particular selective pressure hints at the breathtaking complexity of selective environments.

To appreciate this great complexity, imagine trying to sketch out your own selective environment. You could start with local pressures (*Did I cook that chicken well enough?*), then reach out farther (*Is there a contaminant in my city's water supply?*), but how far do you go? Do you include the properties of the sunlight that come to us from ninety-three million miles away? You should, since it's the source of most of the energy in the biosphere, and also because in years of intense solar flare activity (such as the eleven-year cycle that began in 2005), solar radiation doses can be several hundred times higher than in normal years, "slow-cooking" your DNA.[11] How do you measure the effects of the microbes that may swarm in the water you drink, microbes that might attack an even slightly depleted immune system? What about the *E. coli* bacteria that multiply in your innards every day? Most aren't harmful, but it just takes one harmful strain to arise, spread to your brain, and produce meningitis, which can kill you in just hours. This normally happens in infants, infants who have no idea that such a selective pressure is upon them. Selective pressures don't care if you're an infant or the pope.

Well, obviously you could drive yourself crazy trying to sketch out your selective environment, and contemplating it can even lead to paranoia. By his fifties, eccentric billionaire Howard Hughes tried to shield himself from selective pressures—namely, germs—by insisting on handling everything with clean paper towels. Maybe he wasn't so crazy after all?[12] Can we hide? Can we isolate ourselves?

It turns out that although we can use our technologies to buffer out some selective pressures (such as wearing clothing to keep us warm), we can't protect ourselves from *everything*.[13] Nothing, it turns out, lives in a vacuum, unaffected by the rest of the universe. Selective environments are so complex that we may as well consider them infinite.

Consider life ten thousand feet below the surface of the ocean, where the water is nearly freezing and over two tons of pressure press on every square inch of whatever ventures this far down.[14] Except for the ghostly flashes of occasionally glowing fish, there is no light here:

sunlight doesn't reach this deep. But there is life. There are beard worms, red-and-white "living tubes" standing up to ten feet tall, and there are free-floating colonies of bacteria, feeding on sulfurs billowing up from hot-water vents in the seafloor. And a few fish quietly glide by. But it looks bleak to us humans, like the surface of the moon, and it's hard to escape the notion that the beard worm is among the most isolated, lonely forms of life. But even here, there is a selective environment, and that means selective pressures for the beard worm to endure. Beard worms can't stand the cold water just a few feet from the hydrothermal vents, so they need a way of reproducing that keeps them close to the warmth. Their environment is also affected by a steady rain of decaying organic matter—disintegrated plant and fish bits from far above—that descends like slow snowflakes, and affects the chemistry of the seafloor, in which many beard worms are half-buried. Beard worms battle no one in their cold, dark domain, but they still have a selective environment.

No life, then, is an island unto itself. Each individual is immersed in a complex web of selective pressures that requires environment-specific "solutions." No single adaptation assures fitness in every time and place. And this is exactly why geneticists consider gene-pool diversity to be a measure of health in a population; if everyone is identical, a single environmental change, for example, or a single disease, could devastate the whole population. Genetic diversity is genetic health, a hedge against catastrophe.

## Do the Fittest Survive?

By taking apart the phrase "survival of the fittest," we've been able to see how it masks some important complexity. Yes, it's the fittest that survive. But in an enormously complex world of ever-changing selective pressures, no single characteristic—such as brute strength—ensures survival in every circumstance. What it is to be "fit" depends on what you are, where you are, and when.

## Notes

1. British poet Alfred Lord Tennyson (1809–1892), whose poetic genius leaned toward morbid and violent themes, characterized nature as "red in tooth and claw" in an elegy published nearly a decade before Darwin published *On the Origin of Species*. See M. Demoor, "His Way is thro' Chaos and the Bottomless and Pathless: The Gender of Madness in Alfred Tennyson's Poetry," *Neophilogus* 86, no. 2 (2002): 325–35.

2. But plants deserve some television airtime: most life is supported by plants, which produce both food we can eat and oxygen we can breathe. They also "scrub" our atmosphere of carbon dioxide, which we can't breathe. We emit carbon dioxide every time we exhale and every time we start a car.

3. The popular focus on competition, among economists, biologists, and sociologists alike, is rooted in mistakenly attributing purposive, violent natural competition to Darwin, while this view actually predates Darwin. See note 1 above.

4. See C. Darwin, *On the Origin of Species*, 6th ed. (London: Murray, 1872), p. 63, http://pages.britishlibrary.net/charles.darwin/texts/origin_6th/origin6th.fm_html (accessed December 10, 2005). Although Darwin says "destruction," he means it in a sterile, Victorian way, rather than implying purposive violence or combat. Darwin first used the phrase "survival of the fittest" in the title to chapter 4 ("Natural Selection; or The Survival of the Fittest") of the sixth edition (considered the definitive edition) of *On the Origin of Species*, published in 1872, thirteen years after the first edition. He took the phrase from Herbert Spencer, an early social Darwinist.

5. For a thorough discussion of fitness, see S. Wright, *Evolution and the Genetics of Populations: A Treatise* (Chicago: University of Chicago Press, 1968). A more popular account is found in R. Dawkins, *The Selfish Gene* (New York: Oxford University Press, 1976).

6. Discrete, easily defined selective pressures, such as a major food competitor or a predator, are known as *selective agents*. For many examples of natural selection, see J. Endler, *Natural Selection in the Wild* (Princeton, NJ: Princeton University Press, 1986).

7. Neanderthals were "protohumans" that lived in Europe and the Near East from about two hundred thousand years ago to about thirty thousand years ago. We share a common ancestor with them, but they went

extinct, while we "modern" humans survived. Neanderthal life was not apparently as brutish as Hollywood makes it seem, but it was fundamentally different from the life of "modern" humans, even one hundred thousand years ago. This sketch of Neanderthal life is compiled from O. Bar-Yosef, "Eat What Is There: Hunting and Gathering in the World of Neanderthals and Their Neighbors," *International Journal of Osteoarchaeology* 14, nos. 3–4 (2004): 333–42; K. V. Boyle, "Reconstructing Middle Paleolithic Subsistence Strategies in the South of France," *International Journal of Osteoarchaeology* 10, no. 5 (2000): 336–56; C. Gamble, *The Palaeolithic Societies of Europe* (Cambridge: Cambridge University Press, 1999); and E. Trinkaus and M. Zimmerman, "Trauma among the Shanidar Neanderthals," *American Journal of Physical Anthropology* 57, no. 1 (1982): 61–76.

8. As we will see in "Myth Two: It's Just a Theory," sexually reproducing species combine the male and female DNA, creating offspring slightly different from their parents. Clones are more common in asexually reproducing species (which produce legions of offspring, like Xerox copies) but even so, there is variation. Note that some slime molds (including the Dog Vomit slime mold, which looks just like it sounds) reproduce with an unusual form of sexual reproduction. See S. L. Stephensen and H. Stempen, *Myxomycetes: A Handbook of Slime Molds* (Portland, OR: Timber, 2000).

9. For many examples of selection in the natural world, see J. Endler, *Natural Selection in the Wild.*

10. See M. Menotti-Raymond and S. J. O'Brien, "Dating the Genetic Bottleneck of the African Cheetah," *Proceedings of the National Academy of Sciences USA* 90, no. 8 (1993): 3172–76. See also S. J. O'Brien et al., "Genetic Basis for Species Vulnerability in the Cheetah," *Science* 227, no. 4693 (1985): 1428–34.

11. Solar Physics Group (NASA), "Sun Facts," http://science.msfc.nasa.gov/ssl/pad/solar/sunspots.htm (accessed August 24, 2005).

12. Hughes may have been somewhat fanatic, but today the Howard Hughes Medical Institute promotes and funds (to the tune of over $10 billion a year) a wide variety of important biomedical research projects.

13. Although it may seem that our layers of cultural practices (such as incest avoidance) and, especially, technologies (such as modern medicine) have totally buffered humanity from selective pressures, this is partly an illusion. American fetuses, for example, have recently been found to be

floating in the womb in an unhealthy broth containing mercury, pesticides, and Teflon; in this case, we're *creating our own selective pressures* by polluting our environment! See the nonprofit Environmental Working Group's report "Bodyburden: The Pollution in Newborns" (which has prompted some members of Congress to tighten pollution regulations), http://www.ewg.org/reports/bodyburden2/ (accessed December 22, 2005).

14. For an overview of the fascinating world of deep-sea biology, see J. D. Gage and P. A. Tyler, *Deep-Sea Biology: A Natural History of Organisms at the Deep-Sea Floor* (Cambridge: Cambridge University Press, 1996).

## *Myth Two*

# It's Just a Theory

Have you ever heard someone dismiss evolution by saying that it's just a theory? This view is so common in the United States that even many who accept evolution think of it as just a theory, as though it lacked scientific support. And some believe that if evolution is just a theory, then other "theories," namely, creationism, or its dressed-up twin, intelligent design, should be taught alongside evolution in public schools. In a number of American states, school boards have felt pressure to do just that. Where this has failed, some opponents of evolution have tried to require biology teachers to teach their students that evolution is just a theory.[1] In a recent attempt to dismiss evolution as just a theory, Georgia's Cobb County School District moved to put stickers on high school biology textbooks, stating that:

> This textbook contains material on evolution. Evolution is a theory, not a fact, regarding the origin of living things. This material should be approached with an open mind, studied carefully, and critically considered.

Who can dispute that all scientific work should be studied carefully and critically considered? And having an open mind is also an asset, but not so open that your brains fall out. Yet this is exactly what seems to have happened in Cobb County.[2]

The main problem with the textbook disclaimer has to do with two different meanings of the word *theory*.[3] In popular speech, *theory* means a guess or a hunch that can be just as good as any other guess or hunch, as when someone *theorizes* that a light streaking across the night sky must be an alien spacecraft. When scientists use the word *theory*, however, they're referring to "a logical, tested, well-supported explanation for a great variety of facts."[4] Scientific theories are not guesses.

Scientific disciplines don't all use identical methods, but they share a common approach. First, scientists will observe an object or a process (for example, Mars appears to travel backward in the sky about every twenty-six months, cholera rates are higher in poorer countries, there is a great diversity of species on Earth, and so on). They then formulate a question, asking how the object or the process works, or how it came to be. Then scientists try to answer the question in the form of an educated guess, a *hypothesis*. This hypothesis is tested by making a prediction and seeing whether the results of an experiment or further observation fit with the prediction. If the results don't fit, the hypothesis is either rejected, or it's modified and tested again. If the hypothesis eventually fits with the prediction, the scientists share this information—usually in a peer-reviewed journal—so that other scientists can test the hypothesis themselves.[5] Only after a hypothesis repeatedly accounts for the facts or the observations can it then be considered a theory.

And even theories aren't etched in stone. They can be modified. Sometimes they can even be overthrown completely—as with the theories of *geocentrism* and *phlogiston*—but this requires very compelling evidence.[6] What makes the theory of evolution particularly powerful is that its supporting evidence comes from a wide array of individual scientific fields, including biology, botany, ecology, genetics, geology, paleontology, archaeology, embryology, and zoology.

Evolution is a legitimate scientific theory. It's well supported in the same way that the geological theory of plate tectonics supports and explains earthquakes, tsunamis, continental drift, and mountain formation. And it's as well supported as cosmological theories about how planets are formed or the age of our solar system. Some suggest that evolution is flawed because it can't be observed in the laboratory. It's true that evolutionary science doesn't rely as much on lab experiments as chemistry and physics do (although molecular biology and genetics are changing that), but neither do geologists, paleontologists, or other scientists who study historical processes. These historical sciences don't always apply the same methods as chemistry and physics. But we shouldn't expect them to, and we'd be foolish to dismiss them for it. Some complain that since evolution is concerned with the past, it can't make predictions because predictions are about the future. Thus, it's argued, evolution is not a legitimate scientific theory.

Odd as it may sound, predictions don't have to be about the future. Imagine that you're a molecular biologist, and you've observed that humans look more like chimpanzees than any other species. In light of evolution, this would suggest that we not only share a common ancestor with chimps, but also that our common ancestor may be more recent than any we share with other living animals. You know that the more closely related two individuals are, the more they have in common genetically, just as we're genetically more like our first cousins than our second cousins. So you can make a prediction (a hypothesis) that we have more in common genetically with chimps than with any other living species. If this is true, it will definitely support your hypothesis. Indeed, after studying the DNA, your prediction turns out to be correct.[7] Over 95 percent of our DNA is identical to chimpanzee DNA.[8] As a molecular biologist, you would also know that you can tell roughly how far back in time two individuals share a common ancestor by the differences in their DNA.[9] In the case of humans and chimpanzees, that shared ancestor lived between five and six million years ago.

Evolutionary theory also tells us that because species can change drastically over long periods of time, life in the past was different from life today. We can make a prediction that the further back in time we look, the greater the differences will be between older life-forms and modern life-forms. This turns out to be the case. Recent fossils resemble modern life-forms more than older fossils do.

So, predictions *can* be made about the past, and the results of these historical experiments provide compelling evidence supporting evolution. But as we'll see, earlier explanations for the Earth's great variety of life have been forced aside because they're not as well supported as modern evolutionary theory.

## LAMARCKISM

About fifty years before Charles Darwin (1809–1882) published *On the Origin of Species*, a view of evolution we now call *Lamarckism* was proposed by the French naturalist Jean-Baptiste Lamarck (1744–1829). Lamarck was unique for his time in that he tried to apply strictly scientific explanations to the workings of the natural world. Darwin acknowledged Lamarck's scientific approach when he wrote, "He first did the eminent service of arousing attention to the probability of all change in the organic, as well as in the inorganic world, being the result of law, and not of miraculous interposition."[10] Unfortunately, Lamarck's ideas never became popular during his lifetime, and he died blind and in poverty. He was even buried in a rented grave, and after five years his remains were removed and have been lost ever since. Although Lamarck's explanations weren't completely original,[11] he held a number of views that are now rejected by modern evolutionary theory, also called *neo-Darwinism*.[12] Since many people confuse Lamarckism with neo-Darwinism, it's important to understand how they're different.

Lamarck proposed two principles for how evolution works that are no longer supported by the evidence. One is the law of *use and disuse*. The other is the law of *the inheritance of acquired characteristics*.[13] The first law states that the more you use an organ or body part in making your living, the larger and stronger that part will become. The less you use that body part, the smaller and weaker it will become, eventually withering away if not used at all.

The first law makes sense when we're talking about muscles, which become larger the more they're used, and can atrophy if not used at all. But this isn't the case with many other body parts, such as sense organs. Your hearing does not become better the more you use your ears. And the same goes for using your eyes (although you can strengthen the muscles around your eyes by moving your eyeballs, this won't help with nearsightedness or farsightedness despite the claims of those peddling eye exercise kits).

The second law—*the inheritance of acquired characteristics*—states that the characteristics of stronger or weaker body parts acquired during your lifetime will be passed on through reproduction to your offspring. In other words, if you're a hardworking blacksmith, you'll develop strong muscles, and then your children will be born with larger-than-normal muscles because of your activity. If your children are also blacksmiths, they'll pass on even larger and stronger muscles to their children, and onward through the generations.

Lamarck's best-known example of acquired characteristics involves the evolution of the long necks of giraffes. Lamarck speculated that the ancestors of giraffes had shorter necks, and they would stretch their necks to reach high leaves in trees. Their descendants then inherited longer necks because of the stretching of the parents' necks. Lamarck thought that these changes in offspring could lead to a new species in just a few generations.

Unfortunately for Lamarckism, there's no evidence that acquired characteristics are passed on.[14] Lamarckism seems intuitively obvious, but today we know that alterations in our bodies acquired during our lifetime, such as larger muscles or stretched necks, are not encoded in our genes, and so they're not passed on to the next generation. Lamarck's ideas were reasonable for their time, but they turned out to be wrong. His ideas have been replaced by modern evolutionary theory, which better explains how characteristics are transmitted through inheritance, and how new characteristics arise.

## WHAT EVOLUTION IS

Darwin described evolution as "descent with modification." This simply means changes in the properties of organisms over generations. These changes are explained by at least three independent processes that when taken together form what we mean by evolution.[15] These are *replication*, *variation*, and *selection*, and they are all observable facts. Replication is simply reproduction. Variation is genetic differences between parents and their offspring. And selection refers to natural selection, the process whereby those best adapted to their environment tend to survive and pass on their genes to the next generation.

## REPLICATION

Replication, or reproduction, can be either asexual or sexual. Asexual reproduction happens when offspring are created from a single parent without mixing in the genes from a second parent. These offspring are usually identical to that one parent, kind of like Xerox copies. They're natural clones. This form of reproduction is more common in plants than in animals, and it's also how bacteria reproduce. In contrast, sexual reproduction involves *combining* genes from two parents (male and female) to produce offspring. This is how most animals reproduce, as do many plants—through pollination.

## VARIATION

Variation can arise in a number of ways. One of these is recombination. In sexual reproduction, recombination involves the "shuffling" of the various genes in the male and female sex cells (sperm and egg cells in animals) after these cells unite. This "shuffling," which occurs after conception, creates new gene combinations, making the offspring different from their parents. Another way for variation to arise is through mutation. Mutations are rare changes in the genes, which are often inherited. These changes usually occur as copying errors when cells multiply in the early stages of reproduction. Mutations can also be triggered by certain kinds of radiation, chemicals, and viruses. Many mutations are harmful, some are beneficial, but most seem to be neutral. Whether mutations are harmful, beneficial, or neutral may depend a lot on the environment.

Your body has natural defense systems that kill harmful bacteria. But some bacteria have developed mutations that give them resistance to your natural defenses, allowing them to thrive in their environment, that is, your body. In order to kill them, your bodily environment needs to change so that those mutations can no longer provide an advantage to the bacteria. This is where modern medicine comes in. If you had a serious bacterial infection, such as tuberculosis, your doctor would prescribe antibiotics designed to kill the TB bacteria. If you're fortunate, the bacteria will be killed. But bacteria reproduce rapidly—some as often as every twenty minutes. And often a few will have a mutation that provides resistance to a particular antibiotic. The mutation would be beneficial for the TB bacteria—but obviously not for you. Those antibiotic-resistant bacteria will reproduce, passing on their resistance to the next generation, at which point you would need a different (or stronger) antibiotic to kill the bacteria. And of course there may be some bacteria with a resistance to the new antibiotic, and then those mutant bacteria will come to dominate, and then you'll need a new antibiotic—assuming one has been developed—and on and on.[16] Some bacteria, such as TB, are highly infectious, and new resistant strains can spread easily. If an effective antibiotic has not yet been developed to fight these new strains, this could spell trouble. Doctors will tell you to finish the whole supply of antibiotics, even if you're feeling better, because hitting the bacteria with the prolonged, full dose will increase the chances of killing even those with a slight resistance.

Clearly, beneficial mutations play an important role in evolution since they're the ultimate source of genetic change. These variations arise randomly, in the sense that they don't look ahead and plan what will benefit the individual plant or animal.[17] They have no plan at all.

## Selection

Natural selection is the great testing ground of variation. It's the mechanism that chooses which individuals will survive long enough to reproduce and transmit their genes to the next generation. Of course, natural selection doesn't choose intentionally. If a certain variation provides an advantage, then the individual with that variation stands a better chance of surviving. If it survives long enough to reproduce, it will transmit that beneficial variation to its offspring.

Consider the Komodo dragon, that giant reptile that can grow up to nine feet long and weigh over two hundred pounds. They populate three small islands in Indonesia, and they're rapacious hunters, with wild boars being one of their favorite foods. Imagine that a Komodo dragon is hiding in wait in the forest as a wild boar unwittingly approaches. Let's consider that this particular boar has been born with a slightly better sense of smell than his fellow boars. Now just before the Komodo dragon springs to ambush, the boar gets a whiff of it, turns tail, and escapes to safety. Now imagine another boar that was not born with as good a sense of smell as our first boar. He doesn't sniff out the Komodo dragon, and gets pounced on, killed, and devoured by the Komodo, bones, hooves, and all. The boar born with the better sense of smell has an advantage over the other boar, and he stands a better chance of surviving long enough to procreate and pass on his genes for a sharper sense of smell to his offspring. After a period of time the genes for a sharper sense of smell will spread throughout the population because that boar, and others like him, are more likely to survive and reproduce.

The pressures affecting survival can sometimes be so great that the smallest beneficial variations can make a big difference. If a certain variation provides a disadvantage, such as a boar having a slightly worse sense of smell than its fellows, then that boar has a greater chance of becoming dinner for a predator before it has a chance to reproduce and pass on the genes for a slightly worse sense of smell.

It's important to understand that natural selection is not random. It doesn't allow just any variation to survive. In a sense natural selection is negative, in that it culls the herd. And this culling, whether in terms of pressure from predators, temperature changes in the environment, availability of food and water, and so on, will determine which variations survive. But natural selection is also positive in that it allows for variations to accumulate over time. Variations that provide for a slightly better sense of smell can arise many times, with each variation slightly sharpening that sense. Meanwhile, as the boars are evolving, so are the Komodo dragons. They may become faster, or develop better hearing, which makes them better at killing boars. After hundreds or thousands of generations of this escalating arms race, both Komodo dragons and wild boars will have changed significantly from their ancestors.

Another form of selection,[18] called *sexual selection*, can also play a role in the evolution of animals. Sexual selection involves members of one sex preferring certain characteristics in the opposite sex, and then choosing a mate based on which best displays those characteristics.

The peacock's tail feathers are an extreme example of sexual selection at work. Peahens (female peacocks) prefer peacocks with large, colorful tail feathers. The larger and the more colorful a peacock's tail feathers are, the greater his chances of being selected as a mate by a peahen. He will then pass on his genes for larger, colorful tail feathers to his offspring.

Sexual selection can be beneficial, although the selecting is not done with benefit in mind. It takes some energy for a peacock to parade around, displaying his large tail feathers during courtship. A healthier peacock has shinier feathers, and puts on a better show during his courtship display than a weaker or sickly peacock, whose feathers might be crawling with parasites and whose movements are less vigorous. So if the characteristics the female chooses (large, shiny feathers and vigorous displays) are also signs of good health, then the odds are greater that the offspring will be healthy. But the peahen is not thinking that by mating with a healthier peacock she will have healthier offspring. She just happens to be more attracted to peacocks with larger and shinier tail feathers and more vigorous displays. When she mates with one of these peacocks, the genes for large tail feathers will be expressed in her male offspring, and the genes for preferring larger tail feathers will be expressed in her female offspring.[19]

But sexual selection can also have its drawbacks since it still must pass the test of natural selection. Large and colorful tail feathers are great for attracting peahens, but they're also great for attracting predators since they're so visible from far away. Also, the tail feathers may require too much energy to fly with or to drag around, even when folded up. If a characteristic, such as large tail feathers, puts peacocks at great risk, then they might not live long enough to mate. In that case the peahens will have to settle for peacocks with smaller displays, which will then pass on their genes for smaller tail feathers.

## Speciation

The processes just described—replication, variation, and selection—are occurring all the time in nature, and they're the key processes that collectively are called evolution. They are observable facts that no rational person can deny. Some who admit that these three processes exist still try to deny that evolution occurs. They make a distinction between microevolution (the changes that happen within a species) and macroevolution (the changes by which a new species evolves from an older species), and they deny macroevolution. But it's the same processes at work in both microevolution and macroevolution. The only difference is that macroevolution involves a longer timescale, and so allows many changes to accumulate. Accepting one but not the other is like declaring there are decades but not centuries or millennia.

One way for a new species to evolve is when some members of a species become geographically isolated from others. This can happen when a population of organisms extends its range, or when environmental changes break up the population. A number of things can cause geographical isolation, such as storms carrying birds, insects, or plant seeds to a distant habitat, or rising sea levels at the end of an ice age isolating separate populations of animals and plants on islands previously connected by land bridges. When a single population has been divided into two or more groups, new variations won't be shared by the two groups because they're isolated from each other. And if the environments have different food sources, weather, temperatures, soil conditions, predator and prey species, and so on, then different variations will be selected for by natural selection. Over a long period of time the two separate groups will become different enough from each other that if they were to come together, they wouldn't or couldn't interbreed, and they would be considered different species.

The length of time for a new species to arise can be considerable, and this is perhaps why it's so difficult for us to imagine macroevolution. It's easier to imagine if we look at species that diverged more recently and thus look similar, such as horses, donkeys, and zebras, which diverged about four million years ago. It may require a bit more imagination to conceive of the common ancestor of animals that diverged further back in time and look rather different now, such as whales and hippos. But DNA evidence suggests that hippos may be the closest living relatives to whales, with their most recent common ancestor traced back to about fifty-four million years ago.[20] It may be even harder to grasp that far enough back in time there existed a common ancestor of humans and horses, or even of humans and horseradishes. But all living things are related. You're certainly not a monkey's uncle (or aunt), but you're literally a monkey's cousin. We're all distant cousins, but some are more distant than others.

## Disagreements within the Theory

Evolutionary biologists do sometimes disagree about which hypothesis better explains certain observations. But this doesn't mean that they reject the basic mechanisms of evolution—replication, variation, and selection—described above. There are parallels in other sciences. Geologists, for example, agree that the theory of plate tectonics best explains earthquakes and the movement of land masses. But not all agree on the mechanisms that drive the movement of the plates, especially as they relate to forces working very deep within the earth.[21] But we wouldn't suggest that geologists abandon plate tectonics because of disagreements about specific details. Plate tectonics is a relatively new theory, and it's in the process of being refined. Evolutionary biology is in a similar situation. There are disagreements about certain details, but not about the fact of evolution itself.

For example, one disagreement concerns how exactly to read the DNA clock that tells us how far back two individuals share a common ancestor.[22] These disagreements are being published and examined in peer-reviewed journals, and further research is being done to try to sort out these differences.[23] However, all of the debaters agree that evolution occurs.

Another disagreement concerns how humans became relatively hairless.[24] One hypothesis suggests that as our early human ancestors moved from the forests in central Africa to the hot, open savannah, hairlessness kept their bodies cooler, allowing them to forage for food for longer periods of time.[25] Another hypothesis holds that hairlessness provided an advantage against fleas and other biting parasites. Eventually freedom from parasites became a desirable trait in a mate, and hairlessness was a good way to advertise this.[26] Again, it's important to point out that there are disagreements over particular hypotheses within evolutionary science, yet these differences don't lead scientists to reject the theory of evolution. In fact, these hypotheses only make sense within the framework of evolutionary theory.

Science is inherently conservative, in that scientists don't usually embrace alternative explanations at the drop of a hat. They typically demand very convincing evidence before replacing one view with another. But when that evidence has been accumulated, scientific theories can be modified or even abandoned. The peer-review process helps keep science honest because it allows scientists to scrutinize one another's work and to test new hypotheses for themselves. It would take a lot of very convincing evidence to overturn the theory of evolution, and as things now stand, this convincing evidence is not to be found. There's no better explanation for Earth's great variety of life than the theory of evolution, and to dismiss it as "just a theory" is not only incorrect, it's also irresponsible.

## Is Evolution Just a Theory?

We've seen that calling evolution "just a theory" involves a misunderstanding of what a scientific theory is. Evolution is a fact, and the three main processes that make up evolution—replication, variation, and selection—are observable and undeniable. While scientists do sometimes disagree about specific hypotheses within evolutionary theory, they don't reject the theory of evolution itself. Indeed, the theory of evolution does such a good job of explaining so many observations that the biologist T. Dobzhansky wrote, "Nothing in biology makes sense except in the light of evolution."[27]

## Notes

1. M. Matsumura, "Tennessee Upset: 'Monkey Bill' Law Defeated," *NCSE Reports* 15, no. 4 (1995): 6–7; E. Scott, "State of Alabama Distorts Science, Evolution," *NCSE Reports* 15, no. 4 (1995): 10–11.

2. The attempt to place stickers on biology textbooks led to a lawsuit, *Selman v. Cobb County School District*. On January 13, 2005, a federal judge found the sticker policy unconstitutional.

3. Another problem with the textbook sticker incident is the ambiguous claim that evolution is a theory about the origin of living things. This could mean the origins of species by way of ancestral species—which is a fair description of what evolution is about—or this could mean the beginning of life on Earth. Although this second meaning is of interest to evolutionary science, it's hardly accurate to claim that this is what evolution is about.

4. National Center for Science Education, "What's Wrong with 'Theory Not Fact' Resolutions," National Center for Science Education, December 7, 2000, http://www.ncseweb.org/resources/articles/8643_whats_wrong_with_theory_not_12_7_2000.asp (accessed June 12, 2005).

5. Peer-reviewed journals are scholarly magazines where submitted articles are examined for accuracy by experts in a particular field before being published. In order to prevent bias, the reviewing is normally "blind," in that the reviewers are not told who has authored the articles.

6. *Geocentrism* is the theory that Earth is the center around which the sun and the planets revolve. This theory was replaced by *heliocentrism*, which has all the planets including Earth revolving around the sun. *Phlogiston theory* holds that all flammable materials contain an odorless, colorless, tasteless, weightless substance called *phlogiston*, which is given off during combustion. This theory was replaced by the "oxygen theory," which showed that oxygen is responsible for combustion.

7. DNA (deoxyribonucleic acid) is the molecule found in the nucleus

of cells that carries the instructions (or blueprints) for the growth and the development of most living organisms.

8. D. E. Wildman et al., "Implications of Natural Selection in Shaping 99.4% Nonsynonymous DNA Identity between Humans and Chimpanzees: Enlarging Genus *Homo*," *Proceedings of the National Academy of Sciences USA* 100 (2003): 7181–88.

9. The type of DNA used to estimate when the common ancestor of two species lived is *mitochondrial DNA*. Mitochondria are parts of cells that generate fuel for the activity of the cells, and they have their own DNA because (scientists believe) their ancestors were once separate organisms that became incorporated into plant and animal cells. Mitochondria are passed down only from the mother through reproduction, and their rate of change, or mutation rate, is slow and fairly predictable, making them excellent genetic clocks. See note 23.

10. C. Darwin, *The Origin of Species by Means of Natural Selection*, 6th ed. (London: Murray, 1872), p. xiv, http://pages.britishlibrary.net/charles.darwin/texts/origin_6th/origin6th_fm.html (accessed June 14, 2005).

11. For an insightful essay on how Lamarck is frequently misrepresented in history textbooks, see M. T. Ghiselin, "The Imaginary Lamarck: A Look at Bogus 'History' in Schoolbooks," *Textbook Letter*, September/October 1994, http://www.textbookleague.org/54marck.htm (accessed June 18, 2005).

12. Developed since the 1930s, neo-Darwinism (also called the *modern synthesis*) integrates Darwin's theory of natural selection with the theory of genetic inheritance first proposed by Gregor Mendel (1822–1884) and subsequently refined by later biologists. See E. Mayr and W. B. Provine, eds., *The Evolutionary Synthesis* (Cambridge, MA: Harvard University Press, 1980).

13. J. B. Lamarck, *Zoological Philosophy*, trans. H. Elliot (Chicago: University of Chicago Press, 1984), p. 113.

14. Because the field of genetics had not been developed in Darwin's time, he considered the inheritance of acquired characteristics a reasonable view. His hypothesis of *pangenesis* stated that certain traits acquired during the lifetime of an individual, such as large muscles, could be inherited by its offspring. His idea was that each part of the body produces tiny particles called *gemmules*, which enter into the reproductive organs, enabling them to be inherited by future offspring. Darwin acknowledged that his idea of pangenesis was pure speculation and that if it turned out to be mistaken, it

would not refute his position on natural selection. See C. Darwin, *The Variation of Animals and Plants under Domestication*, 2nd ed., vol. 2 (New York: Appleton, 1883), pp. 349–99, http://pages.britishlibrary.net/charles.darwin/texts/variation/variation_fm1.html (accessed June 20, 2005).

15. Other mechanisms affecting evolution include genetic drift and gene flow. For a concise explanation of these mechanisms, see E. Mayr, *What Evolution Is* (New York: Basic Books, 2001), pp. 98–99.

16. Many antibiotic-resistant strains of bacteria—known as *superbugs*—have evolved due to our overreliance on antibiotics, and they can pose a serious health risk.

17. R. Dawkins, *The Blind Watchmaker: Why the Evidence of Evolution Reveals a Universe without Design* (New York: Norton, 1986), p. 309.

18. Other types of selection not discussed here are artificial selection and kin selection.

19. Dawkins, *The Blind Watchmaker*, p. 203.

20. B. M. Ursing and U. Arnason, "Analyses of Mitochondrial Genomes Strongly Support a Hippopotamus-Whale Clade," *Proceedings of the Royal Society of London B* 265 (1998): 2251–55; M. Nikaido, A. P. Rooney, and N. Okada, "Phylogenetic Relationships among Cetartiodactyls Based on Insertions of Short and Long Interspersed Elements: Hippopotamuses Are the Closest Extant Relatives of Whales," *Proceedings of the National Academy of Sciences USA* 96 (1999): 10261–66.

21. W. J. Kious and R. I. Tilling, *The Dynamic Earth: The Story of Plate Tectonics* (Washington, DC: US Government Printing Office, 1996), pp. 53–55, http:// pubs.usgs.gov/publications/text/dynamic.pdf (accessed June 22, 2005).

22. See note 9.

23. For developing views on interpretation of the mitochondrial DNA clock, see A. Gibbons, "Calibrating the Mitochondrial Clock," *Science* 279, no. 5347 (1998): 28–29. See also R. Ota and D. Penny, "Estimating Changes in Mutational Mechanisms of Evolution," *Journal of Molecular Evolution* 57 (2003): S233–S240. See also S. Y. W. Ho et al., "Time Dependency of Molecular Rate Estimates and Systematic Overestimation of Recent Divergence Times," *Molecular Biology and Evolution* 22 (2005): 1561–68.

24. Of course humans are not truly hairless. The differences in the number of hair follicles between humans and other apes is not that significant. What makes human hair different is that it is short and fine.

25. P. E. Wheeler, "The Evolution of Bipedality and Loss of Functional Body Hair in Humans," *Journal of Human Evolution* 13 (1984): 91–98.

26. M. Pagel and W. Bodmer, "A Naked Ape Would Have Fewer Parasites," *Biology Letters* 270 (2003): S117–S119.

27. T. Dobzhansky, "Nothing in Biology Makes Sense Except in the Light of Evolution," *American Biology Teacher* 35 (1973): 125–29.

## *Myth Three*

# The Ladder of Progress

In 1579 the Franciscan missionary Didacus Valades penned a metaphorical drawing of the ordered levels of all living things from the lowliest life-forms—which he placed at the bottom of his drawing—to the highest life-forms at the top. His drawing shows these levels connected together by a ladderlike chain. Today, many people believe that evolution is similar to Valades' drawing, involving a ladderlike progression, as though nature had a built-in aim to strive ever "upward," rung after rung, from simple to more complex organisms, finally culminating in humans—the ultimate goal of evolution—perched triumphantly at the top rung of the ladder. It's thought that this striving in the direction of perfection (meaning humans, of course) is part of the natural process of evolution.

This idea of evolution or any natural process having an ultimate goal or a purpose is known as *teleology*. But scientists, for all their searching, haven't discovered any evidence of teleology in evolution. There appears to be no inherent drive that propels the evolution of species "upward" toward the ultimate goal of humans, or of any other species. But this idea of higher and lower levels of life has a long history, and perhaps a look at that history will explain why this image is still with us today.

## The Great Chain of Being

All human societies have ways of classifying things and putting them into categories that bring order to the world. During the medieval and Renaissance periods, Europe was no different; there was a grand classification system that was biological, geological, and theological, all in one. This system was known as the Great Chain of Being, or the *Scala Natura* (Nature's Ladder), and it was designed to inventory and categorize everything that existed in the known universe in an orderly way.[1] Like Valades' drawing, it was structured in a ladderlike hierarchy placing the least perfect things at the bottom of the chain (or ladder) to the most perfect at the top. In its simplest form the Great Chain of Being consisted of rocks and minerals at the bottom, followed by plants, then animals, then humans, then angels, and finally God.[2] This gradation of nature, from the least perfect thing on up, was thought to be how God created everything in the universe in an orderly and hierarchical way.

Each of these categories of the chain below God was divided into many smaller sections in order to fit every known being and thing into its proper place. Arthur O. Lovejoy's historical account of the Great Chain of Being reveals its cultural influence:

> The result was the conception of the plan and structure of the world which, through the Middle Ages and down to the late eighteenth century, many philosophers, most men of science, and, indeed, most educated men, were to accept without question—the conception of the universe as a 'Great Chain of Being,' composed of an immense . . . number of links ranging in hierarchical order from the meagerest kind of existents . . . through 'every possible' grade up to the *ens perfectissimum* [most perfect being, i.e., God].[3]

Generally, within the lowest link, rocks of no value to humans were at the bottom, fertile soil was higher up, and various gems and metals found in the earth were even higher, with gold (or diamonds) at the top. Within the plant category, the higher ranking usually went to plants beneficial to humans, such as fruit trees, or to those considered beautiful, such as roses; weeds and poisonous plants were at the bottom. In the animal category, those deemed more noble or intelligent were usually ranked higher than pests or dangerous animals. The human category ranked people according to their social station in life, with kings and popes high above lowly peasants.[4] Even among angels, the archangels ranked higher than ordinary angels.

This Great Chain of Being scheme, however, was not about evolution, an idea that didn't exist in any intelligible form until the late 1700s. There was no notion at this time of new species evolving from older species, or of all species being related and sharing a common ancestor, because all living things were seen as *fixed*, in that they were all created by God during the time of creation and in the same form then that they have now. The Great Chain of Being simply ranked everything in its proper place, in an orderly and rational manner, according to God's original plan. It's no surprise that humans and what humans valued fared quite well in the rankings.

Some later thinkers, such as Lamarck, tried to integrate the Great Chain of Being with the idea of species changing over time. Lamarck thought (correctly) that new species could evolve from older species by natural means, but he believed that evolution moved in a straight, upward direction toward perfection along something resembling the ladder of the Great Chain of Being, with humans at the top, of course. Thus, he thought that humans should be the ultimate benchmark by which all animals are judged, writing:

> It is clear that since the organization of man is the most perfect, it should be regarded as the standard for judging of the perfection or degradation of the other animal organisations.[5]

The Great Chain of Being scheme was so pervasive during Lamarck's time that it made sense to try to link it up with evolution. In a way it wasn't a bad idea considering how little was known about the workings of evolution. And to this day the idea of a natural, goal-directed striving up the ladder, or the chain, in the direction of humans makes for a compelling image that's at the root of many of our misunderstandings about evolution. This image, however, is incompatible with how we now know evolution works, precisely because it's teleological. Not only has no mechanism been found that directs evolution "upward" in a straight line, but also frequent extinctions, and unpredictable changes in the direction in which so many organisms have evolved, show otherwise.

## New and Improved?

If we view evolution as striving upward, it's easy to think that more recent species will be "more evolved" or better adapted to their environments than species with longer histories. Surely we can say that newer species are more fit than older species, and that this is progress. Or is it? It would be a mistake to *assume* that a newer species is better adapted or more fit than an older species just because it's newer. If Earth was hit by a large asteroid, similar to the one that appears to have wiped out the dinosaurs (plus half of all living species) sixty-five million years ago, then just about any living thing would be hard put to survive. That asteroid, which slammed into the earth near the Yucatan Peninsula in Mexico, is estimated to have been ten billion times more powerful than the atomic bomb dropped on Hiroshima.[6] No animal, no matter how new it is, has evolved a thick enough skin or shell to protect it from that kind of devastation. Some fortunate animals might be safe from the firestorms that would quickly engulf the earth after an asteroid impact if they were far enough away, in a naturally protected area. But if firestorms aren't enough to cope with, a massive asteroid smashing into the earth would kick up so much dust and debris that it could block out the sun for months. Before long, plants would die from lack of sunlight, then animals that feed on plants would die, then animals that feed on animals would die. Yet, if a particular species of animal was small enough, could eat most anything, and existed in large enough numbers, then some of them just might be lucky enough to survive the ensuing "nuclear winter" if they could find enough rotting plant and animal matter to feed on.

If such an asteroid were to hit Earth now, older creatures, such as cockroaches and rats, might stand a better chance of avoiding extinction than more recent arrivals, such as humans. While some newer species may have advantages over older species in particular environments, in the face of a massive catastrophe all bets are off. Being a newer species, then, isn't enough to ensure survival.

Catastrophes such as asteroids aside, what ultimately matters is how well adapted the members of a species are to their selective environments. If those environments alter in significant ways, such as in temperature, weather or climate changes, the availability of food and water, the introduction of new predators, and so on, we can't assume that a more recent species will be better adapted than an older species to the new environmental conditions.

The terms *higher* and *lower* add more confusion to our understanding of evolution. These terms simply mark the arrival of a species on the historical tree of life, known as the *phylogenetic tree*. A species that arose further back in time is lower, and one that arose more recently is higher. But these designations don't mean that a species is more or less adapted to its environment.

Chinese giant pandas are higher on the phylogenetic tree than Nile crocodiles, in that their species is a more recent arrival.[7] But a good case could be made that Nile crocodiles are better adapted to their environment because they'll devour just about any animal they can sink their teeth into. Giant pandas, on the other hand, have evolved a very specialized diet, relying almost entirely on bamboo. Depending on only one food source, as giant pandas do, is a risky way to make a living. If the bamboo that the pandas rely on was wiped out by disease or by any other cause, it could spell their demise, whereas Nile crocodiles—having a more varied diet—may well survive if a number of their food sources disappeared. So, the fact that giant pandas are higher on the phylogenetic tree than Nile crocodiles doesn't mean they're better equipped to survive. It just means they're more recent.

## COMPLEXITY

Just as the words *higher* and *lower* can be misleading, so can the notion of complexity. Living things today are, on average, more complex than their ancestors of billions of years past. Some of the earliest life-forms were simple bacteria, and if they were to undergo any changes or adaptations, there'd be no place to go but in the direction of complexity, since you can't get much simpler than bacteria.[8] It's obvious that natural selection has helped shape the complex evolutionary changes through the generations, from single-celled organisms to simple plants and animals, leading to fish, reptiles, amphibians, birds, and eventually mammals. So doesn't this show that progressive complexity is a necessary part of the evolutionary process for all living things? The answer is no.

Bacteria are simple organisms, and many have not changed much for billions of years. Indeed, they may be the most successful group of organisms on the planet, having an estimated total biomass (weight) greater than all other living things combined.[9] Crocodiles haven't changed much either from their days living among the dinosaurs over two hundred million years ago. The rather prehistoric-looking Coelacanth fish, which first appeared around three hundred and fifty million years ago, was thought to have gone extinct sixty-five million years ago. But one was caught alive in 1938 off the south-east coast of Africa, and since then over two hundred have been discovered. These "living fossils" don't appear to have changed much when compared with the fossilized remains of their ancient ancestors.

It would be a mistake, then, to assume as a general rule that complexity is a universal trend in evolution, or that it always confers an advantage for survival. What matters is how well organisms are adapted to their environment, and how well they can adapt to frequent changes in that environment. If evolutionary changes toward complexity provide an advantage for those who acquire them, then those changes will be selected for. If not, they won't. Interestingly, some species lines have become less complex over time, such as cave-dwelling fish that no longer have functioning eyes, and some internal parasites that have lost all means of self-locomotion.[10] Moreover, the skulls of birds and mammals have become simpler than those of their early fish ancestors.[11]

## PROGRESS IN EVOLUTION

So how exactly does the idea of progress apply to evolution? Evolutionary adaptations can build on themselves in a cumulative way, and these adaptations can help improve the way of life for a particular population of organism. These gradual, beneficial changes in lineages can be seen as evolutionary progress.[12] In "Myth Two: It's Just a Theory," we saw how wild boars can develop a slightly sharper sense of smell, allowing them to better sniff out and to escape more readily from Komodo dragons. It was also pointed out that Komodo dragons can evolve advantages themselves, such as becoming faster or acquiring finer hearing, making them better at preying on boars. This continually evolving relationship between predator and prey species is known as an *evolutionary arms race.*[13]

Within this arms race, spanning hundreds or thousands of generations, both the prey and the predator can evolve in a progressive sense. If the wild boars evolve a slightly sharper sense of smell with which they can elude Komodo dragons successfully, then the boars will have an edge in the arms race. If no beneficial variations arise in the Komodo dragons that could help them counter this edge, then the population of Komodo dragons may die of starvation, depending on the availability of other prey species. But a beneficial variation, such as slightly larger leg muscles or better hearing, may arise and be passed down through the generations, eventually spreading throughout the Komodo dragon population precisely because it confers an advantage. Then the Komodo dragons will have the upper hand in the arms race. If the population of boars can't counter this advantage with another beneficial variation of their own, they might all be eaten eventually, and the arms race will come to an end.

But many arms races have continued for some time, with prey getting better at escaping predators, and subsequently, predators getting better at capturing prey, and round and round. There's a kind of ongoing feedback loop between the two species. Those that can't compete in the arms race either starve or get eaten.[14] Those that survive tend to be the ones better adapted to the changing conditions. The kinds of cumulative changes that provide evolutionary improvements in arms races are numerous. They can include improvements in ears, noses, eyes, muscles, teeth, claws, camouflage, communication signals, brain functions, and any other changes that can provide an edge in the arms race.

A spectacular example of an animal evolving variations under pressure to avoid predation is the octopus. Its ability to change its skin color and texture within a second to almost perfectly match its surroundings is so amazing that it seems like a high-tech Hollywood special effect, and it certainly makes the chameleon look like a novice.

So, over hundreds or thousands of generations, Komodo dragons and wild boars (and other animals in predator/prey struggles) may evolve progressive adaptations, precisely because of the arms race between them. Over time, the Komodo dragons and the boars will become much better at hunting and fleeing, respectively, than their ancestors were.[15] If you could travel back in time and return with a wild boar from thousands of generations ago, that boar wouldn't stand a chance against a modern Komodo dragon because the modern Komodo dragon has progressed from the dragons that the boar would be used to.[16] The same goes for transporting an ancestor of the Komodo dragons into the present. They would be hard put to catch any modern boars because the modern boars have also progressed.

Curiously, as this arms race develops, the modern populations of boars are really no better at eluding modern populations of Komodo dragons than their ancestors were in their arms race. Likewise, the modern Komodo dragons are no better at catching modern boars than their ancestors were. Since the dragons and boars are coevolving within this arms race, they're still pretty much evenly matched, despite their being better at escaping or at hunting, respectively, than their ancestors were.

This pattern of evolutionary improvement occurring while still being evenly matched with one's "enemy" is known as the *Red Queen Effect*.[17] This is a reference to a scene in Lewis Carroll's *Through the Looking Glass*, where Alice and the Red Queen are running hand in hand as fast as they can, yet getting nowhere. The trees and other scenery around them haven't changed at all. Feeling confused, Alice says, "Well, in *our* country you'd generally get to somewhere else—if you ran very fast for a long time as we've been doing." "A slow sort of country!" says the queen. "Now, *here*, you see, it takes all the running *you* can do, to keep in the same place. If you want to get somewhere else, you must run at least twice as fast as that!"

So, we can talk about progress in terms of the evolutionary adaptations that build on themselves in a cumulative way, where these adaptations help improve the way of life of members of a species engaged in a predator/prey arms race. This progress isn't measured by how successful the predators or the prey are in their arms race—which doesn't vary much—but by the changes in the equipment for success, such as a keener sense of smell or faster legs, possessed by participants in the arms race.[18] Progressive arms races aren't limited to predator/prey relationships either. They can also exist between parasites and their hosts, and between plants and herbivores.[19]

We can also view progress "purely empirically as the achievement of something that is somehow better, more efficient, and more successful than what preceded it."[20] Natural selection generally favors those individuals better adapted to their environment and usually eliminates those who aren't as well adapted. So, those who survive in a particular lineage are on average better adapted than those who don't survive. Evolution in this sense is definitely progressive.[21]

Throughout the long history of evolution various innovations (or transitions) have arisen, enhancing the adaptability of those organisms that possess them.[22] Such innovations include the origin of chromosomes (parts of a cell that contain genetic instructions), eukaryotes (cells with a clearly defined nucleus), multicellularity, sexual reproduction, specialized organs and structures such as the eye, endothermy (warm-bloodedness), parental care of the young, large central nervous systems, and even language and culture. The organisms in which these progressive innovations first began to appear were quite successful, and this contributed to the spread of their ecological influence.[23] The fact that many of these innovations are shared by large numbers of species attests to their being successful evolutionary improvements.

But these improvements were not inevitable. They didn't have to happen. And these progressive changes were not part of some built-in aim of evolution, nor are they predictable.[24] Furthermore, these evolutionary transitions may not be permanently advantageous over extremely long periods of time, such as a billion years, because it's probable that major environmental catastrophes (such as space debris impacts) will occur over this long time scale, causing extinction no matter how well adapted species are. Similarly, the progressive changes in arms races between specific predator and prey species may last millions of years, but they're unlikely to last hundreds of millions of years.[25] So, progress makes the most sense when talking about gradual changes within a particular evolutionary line during a long, but limited, time period, *not* over the entire history of evolution or the evolution of *all* living things. This becomes obvious when we realize that about 99.99 percent of all evolutionary lines that have ever existed are now extinct, many of which (such as dinosaurs) had surely evolved progressive improvements in their equipment for survival.

## The Measure of Man

We must avoid the temptation to measure all forms of progress in terms of changes that have been improvements in the evolution of humans. Bigger brains, bipedality, opposable thumbs, extended care of the young, and development of language and culture are all features that have been progressive adaptations in the human line. But when we measure progress in other species, we must use criteria that apply to them, not necessarily to us, although we may share some of the same adaptations.

It may help us reduce our human bias if we view various evolutionary changes as progressive when they offer improvements from an engineering perspective in the equipment for survival.[26] Not that engineers couldn't conceive of better "designs," but that they could clearly recognize improvements in efficiency in the same way they could recognize analogous improvements in the transition from the first flying machine of the Wright brothers to the modern jet. New "variations" have been added over time to flying machines that make them faster, safer, and more efficient. And many "variations" have been eliminated, in a manner analogous to natural selection, if they do just the opposite.

Evolutionary changes such as a sharper sense of smell in wild boars; improved echolocation (radar) in bats; strength and dexterity of elephant trunks; bigger teeth and claws in lions; agile wings of swallows; plus various innovations shared by numerous species such as multicellularity, sexual reproduction, and specialized organs, are all examples of progress in terms *relevant to the ways of life of those animals that possess these improvements*. Ultimately, human evolution is not the yardstick by which to measure progressive adaptations in other species.

Humans do seem to be the most intelligent species on Earth, but this wasn't inevitable. Things could easily have turned out otherwise. If the chimplike ancestors of humans had not moved out of the forest and into the savanna, the whole course of human evolution might not have happened. If our early ancestors had been unable to adapt to their new environment, then chimpanzees might be the most intelligent animals on Earth without being much different than they are today.

There are plenty of "what ifs" in the history of the evolution of every species. If certain variations had not arisen, if selective pressures and environments were different, and if fewer or more natural catastrophes had occurred, Earth would surely be a very different place. And even though humans happen to be the most intelligent species on Earth, it was just a matter of chance, not part of a built-in goal of evolution.

## The Big Picture

The image of a Ladder of Nature or Great Chain of Being suggests that evolution has a goal or an overall direction, often thought to be humans. But we know this isn't the case. There is no intent, or ultimate aim, in replication, variation, selection, nor any other mechanism of evolution. Although there have been progressive improvements in various evolutionary lines, and we do recognize the evolution of complexity over time, perhaps the image of a bush works better as an analogy for the *big picture* of evolution. Unlike a ladder or a chain, a bush can branch off in many directions—up, down, left, right, and anywhere in between—and new branches can sprout off of older branches without implying that those farther from the trunk are more perfect or better adapted to their environments than those nearer to the trunk.

## NOTES

1. The Great Chain of Being idea has its roots in ancient Greece, in that its creators borrowed heavily from Aristotle's taxonomy (categorizing of living things) and Plato's idea of the Good. These borrowed ideas were altered to accommodate a Christian worldview. For an enlightening intellectual history of the Great Chain of Being idea, see A. O. Lovejoy, *The Great Chain of Being: A Study of the History of an Idea* (Cambridge, MA: Harvard University Press, 1936).

2. The lowest category of the Great Chain of Being was considered more inclusive than only rocks and minerals. Basically, anything that simply existed (but was not alive) was placed in this category, which also included celestial bodies, fire, and water. The plant category added life to simple existence. The animal category added mobility and passions to existence and life. And the human category included reason on top of existence, life, mobility, and passions.

3. Lovejoy, *The Great Chain of Being*, p. 59.

4. The Great Chain of Being scheme also explained morality as understood by the Christian views of the times. The lowliest forms (minerals, plants, and animals) were considered strictly physical, without spirit. The angels and God were seen as pure spirit. Humans fell somewhere in between, being both physical and spiritual. Individual human moral struggles were seen as battles between the material world through the temptations of the flesh (lust, greed, anger, jealousy, etc.), and the spiritual world through the pull of one's God-given conscience.

5. J. B. Lamarck, *Zoological Philosophy*, trans. H. Elliot (Chicago: University of Chicago Press, 1984), p. 73.

6. For more on the evidence supporting the asteroid impact theory of

dinosaur extinction, see W. Alvarez and F. Asaro, "An Extraterrestrial Impact (Accumulating Evidence Suggests an Asteroid or Comet Caused the Cretaceous Extinction)," *Scientific American* 263, no. 4 (October 1990): 78–84. See also R. Cowen, *History of Life*, 4th ed. (Malden, MA: Blackwell, 2005), chap. 16. The energy released from the asteroid is estimated to have been equivalent to one hundred million megatons of TNT. The Hiroshima bomb was 0.01 megatons.

7. It's estimated that pandas first appeared around fifteen million years ago, whereas crocodiles first appeared around two hundred and forty million years ago.

8. For a detailed explanation of how the simplest life-forms such as bacteria can't really get much simpler, and if they did change it would inevitably be in the direction of complexity, see S. J. Gould, *Full House: The Spread of Excellence from Plato to Darwin* (New York: Harmony Books, 1996), chap. 13. For a critique of Gould's position on the implications of this view as it applies to progress in evolution, see R. Dawkins, "Human Chauvinism," *Evolution* 51, no. 3 (1997): 1015–20.

9. E. Mayr, *What Evolution Is* (New York: Basic Books, 2001), p. 278. See also Gould, *Full House*, p. 194.

10. Dawkins, "Human Chauvinism," p. 1018; Gould, *Full House*, pp. 200–201.

11. Mayr, *What Evolution Is*, p. 214.

12. This definition of progress is a slight simplification of Dawkins's excellent definition of progress, which he defines as "a tendency for lineages to improve cumulatively their adaptive fit to their particular way of life, by increasing the numbers of features which combine together in adaptive complexes." See Dawkins, "Human Chauvinism," p. 1016. Evolutionary lineages (or lines) refer to the sequence of ancestral populations to descendent populations.

13. For a thoroughly convincing explanation of how predator/prey arms races can count as progress in evolution, see R. Dawkins, *The Blind Watchmaker: Why the Evidence of Evolution Reveals a Universe without Design* (New York: Norton, 1986), chap. 7.

14. A population of animals can sometimes participate in more than one predator/prey arms race. Lions, for example, may participate in arms races with zebras, wildebeests, gazelles, wild pigs, and others, so that if a

group of lions cannot match one of their prey species in a progressive arms race, they might not starve since they have other prey species available.

15. Progress within arms races is not necessarily constant. Changing environmental conditions can cause the arms race to halt, or even go "backward" for periods of time. See Dawkins, *The Blind Watchmaker*, p. 181.

16. Dawkins suggests the idea of a time machine in order to illustrate the progressive changes in particular lineages over time. See Dawkins, *The Blind Watchmaker*, p. 183.

17. The term *Red Queen Effect* used to describe these evolutionary changes was coined by the American biologist Leigh Van Valen. See L. Van Valen, "A New Evolutionary Law," *Evolutionary Theory* 1 (1973): 1–30.

18. Dawkins, *The Blind Watchmaker*, p. 183.

19. Arms races between parasites and hosts involve parasites developing better ways to infect their hosts and the hosts developing improvements in thwarting the parasites. Many plant species have evolved better defenses from herbivores, producing bitter-tasting toxic chemicals such as alkaloids, while the herbivores have evolved detoxifying enzymes, overcoming these defenses.

20. *Empirically* used here means from the point of view of an objective, impartial observer. Mayr, *What Evolution Is*, p. 214.

21. Ibid., p. 278.

22. For an account of evolutionary transitions focusing primarily on changes in the method of information transmission, see J. M. Smith and E. Szathmáry, *The Major Transitions in Evolution* (Oxford: Freeman, 1995).

23. Mayr, *What Evolution Is*, p. 215.

24. E. Mayr, *One Long Argument: Charles Darwin and the Genesis of Modern Evolutionary Thought* (Cambridge, MA: Harvard University Press, 1991), p. 65.

25. Dawkins, "Human Chauvinism," pp. 1018–19.

26. Dawkins explains that an engineer would easily recognize the progressive increase in adaptive features in the improvements in the optical quality of the eye, as it evolved from a simple light-detecting organ. See Dawkins, "Human Chauvinism," p. 1018.

# Myth Four

# The Missing Link

"Fossils May Be Humans' Missing Link" reported the *Washington Post* on April 22, 1999, explaining that newly discovered Ethiopian fossils "may well be the long-sought immediate predecessor of human beings." But almost fifty years earlier, paleontologist Robert Broom published *Finding the Missing Link*, a book about his discovery of fossilized "ape men" in a crumbling South African cave. Indeed, the phrase "the missing link" was commonly used even before Broom, and reports of the discovery of "missing links" have been continuous since the 1850s.[1]

You have to ask, "What's going on?" How is it that the missing link has been discovered repeatedly, for at least fifty years, and in places as widely separated as South and East Africa? Are there lots of missing links, or just one? What exactly *is* a missing link, and do anthropologists really spend their lives searching for them? We only really hear about missing links between people and apes, but do other life-forms have missing links? "The missing link" is a common phrase that causes confusion. It also gives rise to two major misconceptions.

First, the phrase leads directly to a profound misunderstanding of the world of living things because it promotes an essentially medieval view of nature, one in which species are fixed "types," or "links," that don't change through time. But as we'll see throughout this book, this view is at spectacular odds with the essence of evolution, which is *change*. Species can't be both changeable and unchangeable over time, so we have a major issue to sort out here.

Second, the phrase leads to misconceptions about the study of ancient life and the people who study it. While evolutionary scientists do sometimes seek missing links, they do a lot more than that, and there's far more involved in the studies of ancient life than the quest for the missing link.

A complicating factor here is that there *are* missing links, but their real significance is often lost in the mass media. Luckily, with a bit of thought, it becomes clear what the term really means.

We have a lot of issues to grapple with here: the myth of single Missing Links, the concepts of changeability and fixity, and misconceptions about what people who study ancient life really do. To clarify what's meant by the common use of "The Missing Link," let's start with the origins of the phrase, and then see how that concept matches up with what we know about biology today.

## First Use of the Phrase "The Missing Link"

It's believed that geologist Charles Lyell (1797–1875) coined this phrase in 1851.[2] For Lyell, a missing link was a fossilized life-form, an intermediate life-form living in the time between the lifetimes of two other known, and apparently related, life-forms. Discussing the significance of Darwin's theory of evolution, Lyell wrote, "newly discovered fossils serve to fill up gaps between . . . types previously familiar to us, supplying often the missing links of the chain, which, if [evolution] is accepted, must once have been continuous."[3]

An example, resting on Lyell's law of superposition, makes this crystal clear. Lyell stated that, all other things being equal, the deeper in the earth a layer, or stratum, is found, the older it is, because layers pile up over time. This simple fact is enormously useful for anyone interested in ancient life, because those layers should preserve a historical record of ancient life through time; each geological layer is a page in an extraordinary book, recounting hundreds of millions of years of life on Earth. Lyell and the early paleontologists showed that we can read this extraordinary tome—the history of life—written in stone.

But the tome, like many relics, isn't complete—it shows the wear and tear of time. Pages, even entire chapters, are occasionally missing. Geological layers were stripped away before being preserved, or thousands of years were crushed into a single, millimeter-thick layer by massive pressures. Such geological actions damaged the history of life, like a book that's been smashed, scorched, and damaged by water and insects. But such damage doesn't make the book entirely unreadable: *of course* there are missing pages in such a book, and *of course* there are missing links in the fossil record.

Reconstructing the history of life is much of what paleontologists do; they're detectives, piecing a story together from a fragmentary record. Imagine that you're one such detective, investigating the Koobi Fora Formation, a geological layer in northern Kenya.[4] Over two million years ago, the sands and the rubble you crawl across were an ancient lakeshore. Deep in the oldest layers, your patient excavations reveal the fossils of protohuman creatures bearing some of the characteristics of modern humans, such as relatively large brains and smallish teeth. But there are no stone tools with them, and that's strange because modern human sites, shallower in the ground, nearly always have stone tools. And their brains, though slightly larger than those of chimps and gorillas, are appreciably smaller than humans.

Most important, fossils of their hips show that they walked upright, and that's very unusual in the primate order. Although it's clear that modern humans and these protohumans are related (many other features of their skeletal anatomy are very similar), the leap from the oldest fossils to modern humans feels like a stretch. We certainly don't see such rapid "leaps" in evolution today. So you keep excavating, hoping to find the missing link between these modern and ancient forms. And one day, excavating a littler higher up the slope from your protohuman discoveries—but below the levels where you find modern human remains—you find the fossils of a third kind of creature: it's neither modern human, nor the ancient form. But it bears so many resemblances to both—it has teeth of an intermediate size, it's found with very crude stone tools, it has a brain smaller than moderns, but larger than the protohuman—that the best explanation is that this is an "intermediate form," otherwise known as the missing link. Of course, the fact that it's no longer missing is exciting, which can lead to getting your research funded for years, so you quickly fax out your press release: "MISSING LINK FOUND!"

The meaning of the phrase "the missing link" then seems to be straightforward. It refers to a group of plants or animals intermediate between known and related forms. We'll see how things get a little more complicated below, but for the moment it's important to note that in the study of evolution there are many, many missing links. This is largely because the basic geological processes that have formed and modified Earth's surface—impact cratering, volcanism, tectonics, and erosion—have destroyed the fossils of countless species, chopping up the record of gradual change. And consider that, for various reasons, not everything that died in the past ended up as a fossil. This has prevented entire sections of the history of life from even being written in the first place.[5] Even if, in some paleontologist's paradise, a planet did form where all life was recorded in perfect fossil layers, it would take that whole planet's population of paleontologists lifetimes to discover and document each and every missing link. That the fossil record is incomplete doesn't undermine the theory of evolution—it's simply a fact of geology.[6]

Nevertheless, here on Earth the fossil record is well enough known that transitional fossils (each a missing link) *have* been discovered between fish and amphibians, amphibians and reptiles, reptiles and mammals, and even "walking whales" and marine whales we know today,[7] as well as many other forms of life, some of which we'll examine below.

But before that, let's clear one thing up first: if the term "the missing link" is so straightforward, how has it led to so much confusion?

## Revisiting the Great Chain of Being

The phrase "the missing link" invokes a metaphorical chain, a set of links that stretches far back in time. Each link represents a single species, a single variety of life. Because each link is connected to two other links, each is intimately connected to its ancestors and its descendants. Break one link, and the pieces of the chain can be separated and the relationships lost. But find a lost link, and you can reassemble the chain by reconnecting the separated lengths. So far, so good; this works for any individual species or lineage of life: there *are* missing links in the fossil record, and we occasionally find them. But the next implication of the missing link phrase starts to lead us into trouble.

People often take the missing link phrase to mean a missing link between humans and all other living things, and that's because many still believe in a hierarchical concept of nature called the Great Chain of Being. As we saw in "Myth Three: The Ladder of Progress," the Great Chain of Being is the idea that each life-form occupies a distinct position in a hierarchy, or a link in the chain, a chain leading from snails to monkeys and right on up to humans. But this is an ancient view, and it's completely wrong.

Rather than *explaining* the evolutionary history and the interrelationships between species (as we do today), early naturalists used the Great Chain of Being to describe God's design of the universe. Significantly, the universe was considered to be largely fixed, or unchanging, rather than dynamic, or changeable (as we understand it today), because it was believed that God had created every type of living thing for a certain, distinct purpose. Describing species allowed one to identify their abilities, and this allowed the discovery of their purposes. Such was the work of many early naturalists, including the Puritan John Ray (1627–1705), a meticulous naturalist whose 1691 tome, *The Wisdom of God Manifested in the Works of Creation*, included a section in which he proposed the functions of all the species he knew.[8] The flavor of Ray's worldview can be found in his comments on the functions of "noxious insects":

> Why there should be so many of them produced . . . I answer . . . that many that are noxious to us, are salutary to other Creatures; and some that are Poison to us, are Food to them. So we see the Poultry-kind feed upon Spiders. Nay, there is scarce any noxious Insect, but one Bird or other eats it . . . [and furthermore] God is Pleased sometimes to make use of [insects, such as locusts] as Scourges, to chastize or punish wicked Persons or Nations, as he did Herod, and the Egyptians.[9]

In Ray's influential concept of the natural world, then, insects were created for various ends: spiders as birdfeed, for example, and locusts as a plague upon the wicked. This view portrays nature as "fixed," with each species of plant or animal designed for a divinely mandated purpose. Locusts were designed for their role by the infallible Creator, and thus have always been the same, and will remain the same. Each species, then, was an unchanging link in the Great Chain of Being.

This model demands two things: First, that forms of life, generally referred to as species, be distinct and easily identifiable, because their form reflects their function. Second, that these forms never change, because the Creator had formed each perfectly for its function from the very beginning.

The missing link concept, then, incorporates two very ancient but easily tested ideas. What does modern biology have to say about them? Are species discrete, easily delineated biological entities? And do they change through time, or remain the same?

## Can You Draw a Line around a Species?

Generally speaking, a species is a group of plants or animals that breeds among its kind, but is so different from other forms of life that it can't breed with them; this is called *reproductive isolation*, and it can happen, for example, when a single species splits into two populations, each moving into a different geographical region and over time adapting so much to those new regions that they no longer (in reproductive terms) "fit."[10] Because the way a plant or animal survives is often clearly indicated by its anatomy, we can often tell different species just by looking at the outward forms; orchids, for example, aren't much like cacti, and polar bears are very different from moles, and the differences between these kinds of life-forms are a reflection of how they "make their living." So far, John Ray's approach seems to be intact.

But only so far. The species concept is actually pretty slippery, and one recent book, dealing specifically with the species concept, lists many different definitions of the word.[11] For example, one researcher might be more interested in defining kinds of life with genetic data, while another researcher is more convinced that the fossil record of that life-form, which isn't subject to genetic testing, is more important. Still others might say that behavior is important to consider, and, of course, some say that all this information needs to be considered. But even living life-forms, for which we have plenty of data, can be hard to draw a line around. For example, lions and tigers once coexisted naturally in India, and although they're outwardly very different, they can mate to create "tigons" or "ligers" in captivity. Since tigons and ligers were never found in nature, it's known that lions and tigers did not interbreed naturally. Genetically, then, lions and tigers can be classified as one species, but behaviorally they differed enough to be considered separate species by biologists, and in nature this difference was maintained by the animals themselves.[12]

What we call two different species might in fact be one, genetically, but two, behaviorally. In this case, two "obviously different" types of life happen to overlap. Another way to think of lions and tigers is as shades in a sort of spectrum of *big-catdom*, shades we might call *liondom* and *tigerdom*. But, like colors, rather than being divided by a gap, and existing as discrete entities, *liondom* and *tigerdom* are gradations on a scale.

And the same applies to any other life-forms, even if it appears, today, to be relatively easy to separate one life-form from any other. To find out why, we have to evaluate the second assertion of the missing link/Great Chain of Being concept—the idea that these links are unchanging.

## Do Species Change or Stay the Same?

By the mid-1800s science widely accepted that Earth was not just a few thousands of years old, but many millions, and perhaps even billions, of years old.[13] Not only that, but the layers of rock documenting this vast sweep of time were known to contain millions of fossils, traces of ancient life, generally arranged in a historical sequence, from earliest to most recent, and it's no wonder that studies of geology and evolution went hand in hand. Geological dating allowed investigators of evolution to evaluate the second implication of the missing link/Great Chain of Being concept of nature: do distinct kinds of life (often known as species) change through time, or do they remain the same?

In Darwin's time there might have been some debate about this, but modern biology is very clear on the point. The characteristics of species do in fact change through time, and we can see these changes in the fossil record, the two-billion-year-old ledger of the comings and goings of millions of plant and animal forms.[14] The best way to see these changes is to look at the characteristics of life-forms in early, middle, and late layers of rock, or sediment, representing the shape of that life-form a long time ago, a little more recently, and, comparatively, very recently.

For example, core samples from layers of seafloor sediment in the equatorial Pacific region contain a detailed stack of layers, samples from sediment at least two million years old. Trapped in the sediment are millions of tiny, cone-shaped plankton called *Rhizosolenia* (they're so numerous that several million are found in every pint of seawater). A recent microscopic study of five thousand of these diatoms from the oldest layers, about three and a half million years old, showed their hyaline area (a part of the cone, near the tip) was around three microns (about .00001 inches, much smaller than the thickness of a human hair) in size. In the middle layers, about three million years old, many of the diatoms have a somewhat smaller hyaline area, although plenty are still around three microns in size. But in the most recent layers, it's clear that there are two different kinds of *Rhizosolenia*; one with about the same hyaline area as the diatoms from three and a half million years ago, but another kind with a hyaline area roughly four times smaller than that of the *Rhizosolenia* from the same time period.[15]

There are two possible explanations for this observation. One is that newly sized diatoms were continually being "popped out," either by some machine or a designer that created finished life-forms and placed them in the sea. This sounds ridiculous, but if life-forms are fixed, as in John Ray's conception, this is the only way to account for this observed change in diatom size through time. Nobody in science believes this, and to believe it would require belief in a machine or a designer for which there is no scientific evidence whatsoever. It's much more reasonable that *Rhizosolenia* itself changed through time.

This, then, is evidence of a change in this life-form through time, and evolution has shown that it applies to all life-forms. It's the reason that the links (including missing links) of a Great Chain of Being just don't exist. Life-forms haven't "popped up," fully formed as links in a Great Chain. And we rarely see radical changes in life-forms, "leaps" from one "link" to a triumphant arrival as "another kind," or as the "next link." Rather, we normally see gradual change through time.[16] Life-forms take their forms slowly, over evolutionary time, as they continuously adapt to environmental conditions, and they even diverge into new groups, which later become so different from their ancestral groups that new species—new kinds of life—are formed.

As we found when we examined the view that species should be easily defined (because of their divinely specified functions), we've now found that the Great Chain of Being just doesn't stand up to modern knowledge. When we look carefully at living things, we find plenty of overlap between them. And when we apply the knowledge that Earth is old and that new forms don't just appear in the fossil record as novel forms, but slowly develop through time, we know *why* life-forms can be difficult to separate: because life is arranged as shades through time.

## What Researchers of Ancient Life Really Do

Early in this chapter we saw that the common perception of paleontologists and archaeologists—or anyone in the business of "looking for old things in the ground" for that matter—is that their life is a quest for the missing link. But as we've seen, actually finding missing links can be very difficult, and it's more productive to simply document the gradual changes in life-forms through time than to obsess over finding them. So why does the myth of the missing link quest persist? It seems to have a lot to do with two factors that have been closely connected for centuries and that are very familiar to us today: self-promotion and drama.

Scientists are human, and they can love the limelight like anyone else. Most would leap at any opportunity to have their findings publicized in the popular media, but few have control over the precise words or phrases used to describe their findings to the public. With regard to studies of ancient life, seeing one's work thrillingly announced as "*Missing Link Found!*" can cause both elation (at the publicity) and a wince (at the archaic phrase). But why does the media care in the first place? Because the search for missing links, for origins, is indeed great drama.[17]

In any search for a fossilized missing link, we can perceive a great drama in the search for our origins, a story fulfilling all six of Aristotle's (384–322 BCE) required elements for drama: the *plot* is the finding of or not finding of the missing link; the *theme* is the heroic quest for that elusive missing link; the *characters* are the strange folk who search the wilderness for fossils; the *dialogue* and *rhythm* are the arcane language of science; and the *spectacle* is that of the unfamiliar—the wild lands where fossils are sought, and the moment of truth: Is the missing link found, or not? Add the fact that the search for fossils of human ancestors is the quest for our own identity, and it's no wonder we've been funding and dramatizing the search for the missing link between humans and our closest living relatives, the chimpanzees, since the earliest days of archaeology and human paleontology.[18] The search for fossils of our ancestors is a great story. Many forget exactly how old the famous early hominid "Lucy" is,[19] but few forget the delicious tidbit that she was named for the Beatles song "Lucy in the Sky with Diamonds" by long-haired fossil hunters in the swingin' '70s!

## Is There a Missing Link?

In the end, then, both the *Washington Post* article ("Fossils May Be Humans' Missing Link" ) and Robert Broom's book (*Finding the Missing Link*) were right. The missing link between humans and chimpanzees, our closest living relatives, *has* been found, time and again, in discoveries of fossils of protohuman creatures that are neither human (of the genus *Homo*) nor chimpanzee (of the genus *Pan*), but of the genus *Aus-*

*tralopithecus*.[20] Australopithecines were African hominids, large primates that walked upright. Their earliest members lived roughly four to six million years ago, not long after chimps and gorillas diverged as forest-adapted primates, and the ancestors of our lineage, the early australopithecines themselves, moved onto the ground. By that time, they were bipedal, like modern humans, but they had large, chimplike teeth, and smallish, chimplike brains. Australopithecines most likely used tools more complex than the chimpanzee's termite-mound probe sticks, but these were far less complex than the symmetrical stone tools made by the early members of our genus, *Homo*.[21] In terms of anatomy and behavior, some australopithecines really do appear to be "half human." And, it's widely believed that early *Homo* descended from some variety of late *Australopithecus*.

But since there were many varieties of late *Australopithecus*, as well as many varieties of early *Homo*, there's no obvious place to draw a discrete line separating a shade of late *Australopithecus* from an early shade of *Homo*.

Clearly, it's more accurate to say that we've found some "shade" between ourselves and the rest of the primates, rather than the missing link, and it's important to realize that the thrilling search for our ancestors is not simply a quest for any missing link, but for the whole spectrum of life that connects us, through time, to every other living thing.

## Notes

1. Even today, the excellent television program *NOVA* aired an entire series on the story of the missing link. This was originally broadcast on PBS in 2002, and aside from the problematic (as we saw in this chapter) title, the series is excellent.

2. The phrase "the missing link" is attributed to Lyell in E. R. Lankester, *Diversions of a Naturalist* (New York: Macmillan, 1915), p. 276.

3. C. Lyell, *The Antiquity of Man* (London: Dent, 1927), pp. 323–24.

Lyell also spoke of chains and links in relation not to biology, but to geology, such as geological layers expected to be intermediate layers between two known layers, but they were "missing" because geological processes had altered or moved them. In this book, however, we're mostly concerned with the concept as it is applied to living things.

4. For an exciting account of the fossil discoveries at Lake Turkana, see R. Leakey and R. Lewin, *People of the Lake* (New York: Anchor, 1978).

5. When paleontologists realized that certain processes could destroy skeletons before they were fossilized, they started an entirely new field of study, the study of the fossilization process, or *taphonomy* (*taphos* meaning "of burial"). The study has been so productive that the field has had its own journal (*Journal of Taphonomy*) since 2003.

6. The lack of missing links, or transitional forms, often demanded by creationists, doesn't mean that the theory of evolution is flawed, only that we don't have a perfect geological record of every species' complete evolutionary history.

7. K. Miller, "Response to Newman," *Perspectives on Science and Christian Faith* 48 (March 1996): 66–68.

8. Like Ray, Carolus Linnaeus (1707–1797), who designed the hierarchical biological classification system we use today (e.g., genus, species, subspecies), was likewise concerned with describing the order of nature, rather than explaining its evolutionary development. Until Lamarck and Darwin broke with the idea that the universe was fixed and unchanging, an evolutionary concept of the universe was hardly even possible.

9. J. Ray, *The Wisdom of God Manifested in the Works of Creation: In Two Parts* (London: Innys, 1717), pp. 374–75. Note that with Ray's functional explanations of insects, the inquiry into the form, nature, and history of the insects comes to a halt. Ray's is a fascinating explanation of locusts and of other creeping things, but the reality behind why life-forms are the way they are is even more fascinating.

10. The question of how to define a species is complex. For example, should we classify life-forms by behavior, superficial appearance, anatomy, genetic characteristics . . . or a combination of these? The definition given in this book is generally applicable. A textbook definition is available in M. W. Strickberger, *Genetics*, 3rd ed. (New York: Macmillan, 1985), pp. 747–56. Fascinating reviews of the species concept are found in J. Mallet, "A Species

Definition for the Modern Synthesis," *Trends in Ecology and Evolution* 10 (1995): 294–99, and E. Mayr, "What Is a Species and What Is Not?" *Philosophy of Science* 63 (1996): 262–77. The classification of life-forms is mentioned again in "Myth Seven: People Come from Monkeys."

11. A recent interdisciplinary review of the species concept is available in R. A. Wilson, ed., *Species: New Interdisciplinary Essays* (Cambridge, MA: MIT Press, 1999).

12. E. O. Wilson, *Sociobiology* (Cambridge, MA: Harvard University Press, 1977), p. 7.

13. For more on the pseudoscientific claims that Earth is only a few thousand years old, see "Myth Eight: Creationism Disproves Evolution."

14. Earth is currently believed to have consolidated into a distinct planet, from cosmic debris, about four and a half billion years ago. A billion is one thousand million. See G. B. Dalrymple, *The Age of the Earth* (Stanford, CA: Stanford University Press, 1991), p. 402. Some microscopic fossil-like features in Australian rocks have been dated to 3.45 billion years ago. See J. W. Schopf and B. M. Packer, "Early Archean (3.3 Billion to 3.5 Billion-Year-Old) Microfossils from Warrawoona Group, Australia," *Science* 237 (1987): 70–73. The earliest widely accepted fossils come from Canada, dated to about two billion years ago. See S. Moorbath, "Palaeobiology: Dating Earliest Life," *Nature* 434 (2005): 155. Even though not all individuals fossilize, the fossil record often contains very complete information on ancient life, providing sufficient data to reconstruct ancient life communities, speciation events, and even individual behavior. For a review see S. M. Kidwell and K. W. Flessa, "The Quality of the Fossil Record: Populations, Species, and Communities," *Annual Review of Earth and Planetary Sciences* 24 (1996): 433–64.

15. For the evidence of change through time, see M. J. Benton and P. N. Pearson, "Speciation in the Fossil Record," *Trends in Ecology and Evolution* 16, no. 7 (2001): 408. The abundance of *Rhizosolenia* in seawater today is estimated in I. Gárate-Lizárraga, S. Beltrones, and V. Maldonado-López, "First Record of a *Rhizosolenia debyana* Bloom in the Gulf of California, Mexico," *Pacific Science* 57, no. 2 (2003): 142, which includes fascinating images of these diatoms.

16. On the timescale of human life, evolutionary changes are usually very slow, even too slow to observe in the 150 years of study since Darwin. It's been said that most life-forms evolve too slow to see, but too fast to fos-

silize. The rate at which evolution occurs is noted again in "Myth Six: People Come from Monkeys." A heated debate in evolution has been whether the rate of speciation is generally slow (*gradualism*), or is interspersed with rapid bursts (*punctuationism*), but we agree with Dawkins that this debate has been exaggerated. See R. Dawkins, "Human Chauvinism," *Evolution* 51, no. 3 (1997): 1018. For more on this debate, see K. Sterelny, *Dawkins vs. Gould: Survival of the Fittest* (Cambridge, UK: Icon Books, 2001).

17. M. Landau, *Narratives of Human Evolution* (New Haven, CT: Yale University Press, 1991) shows how we apply themes of classical drama not only to the search for fossils, but also to the way we reconstruct the story of human evolution.

18. Archaeology and human paleontology were systematized into serious Western disciplines in the mid-nineteenth century. Indeed, Darwin, suggesting humans had a history that went beyond written records, including the Bible itself, opened the door to the study of human *pre*history, where before him, human *pre*history was largely inconceivable.

19. Lucy (specimen AL288-2), the name given to a collection of fossilized bones of a single *australopithecine* (see below), was unearthed in Ethiopia in 1974. She is dated to about 3.2 million years ago. See D. Johanson and M. Taieb, "Plio-Pleistocene Hominid Discoveries in Hadar, Ethiopia," *Nature* 260 (1976): 293–97, or, for a more popular account, D. Johanson and M. A. Edey, *Lucy: The Beginnings of Humankind* (New York: Simon & Schuster, 1981).

20. The genus *Australopithecus* was named in 1923 by Robert Broom. It means Southern (*austral*) Ape (*pithecus*), because it was first found in South Africa, but australopithecines lived throughout sub-Saharan Africa. Australopithecines are divided into two main groups—the *graciles* and the *robusts*. The late graciles appear to have evolved into early *Homo*, while the robusts went extinct around one million years ago. For an overview, see R. G. Klein, *The Human Career* (Chicago: University of Chicago Press, 1999). For more on robust australopithecines, see F. E. Grine, ed., *Evolutionary History of the Robust Australopithecine* (New York: De Gruyter, 1988).

21. Although it remains possible that early *Homo*, and not *Australopithecus*, created the stone implements found with *Australopithecus garhi* (see B. Asfaw et al., "*Australopithecus garhi*: A New Species of Early Hominid from Ethiopia," *Science* 284, no. 5414 [1999]: 629–35), it remains the case

that hominid tool use was far more complex than that of any other animal, and that hominids relied more heavily on such implements than any other creature. It has been suggested that for a number of reasons, complex tool use should not be restricted to the genus *Homo* (see B. Wood and M. Collard, "Is *Homo* Defined by Culture?" *Proceedings of the British Academy of Sciences* 99 [1999]: 11–23). Note also that recent studies support the long-held general impression that australopithecines used simple tools. See L. R. Blackwell and F. D'Errico, "Evidence of Termite Foraging by Swartkrans Early Hominids," *Proceedings of the National Academy of Sciences* USA 98, no. 4 (2001): 1358–63.

## *Myth Five*

# Evolution Is Random

A child pours Legos onto the carpet, making a messy pile of colored blocks, slabs, cylinders, and L-shapes. But as the child snaps them together, order emerges from the chaos. A symmetrical foundation emerges, then walls and towers. Finally, where once there lay just a pile of plastic bits, there now stands a miniature castle, detailed and color coordinated. With great satisfaction, the child has learned that intent can create order from disorder. It is a powerful and profound revelation. From this day forward, the idea that only designers with intent can account for such order is reinforced. How else could the castle come to be? Could the Legos pick themselves up and snap themselves into position? Could random forces, like a tidal wave or a blast of wind, assemble the pieces into a castle? No—that, it seems, demands a designing mind. The order in the universe makes sense to the child that makes a thing; the universe and its living inhabitants, it seems, must also be designed, the products of intent.

But when the child becomes a student, and is first exposed to the concept of evolution, she confronts an apparent paradox. Textbooks state that evolution is the author of the order and the complexity of life, but also that "Evolution is random and undirected."[1] Based on everything the child has experienced, that just doesn't make sense. Thinking back on the obvious order in plants and animals and communities of living things, thinking back to creating order from disorder with her Legos, the student wonders: *Evolution is random? Complexity from randomness? How?* It seems screamingly obvious that the world of living things is the result of an intent or a designer assembling order according to some master plan. Even Einstein said, "I shall never believe that God plays dice with the world,"[2] suggesting a creator, some entity with a plan, or at least that the world and its living things can't possibly be the result of a cosmic game of dice, of random chance.

It's at this point, unfortunately, that many part with evolution. They commonly ask: *How could something as perfect as the eye, or the wing, come from random evolution?* To support their concept of a designer, many often point out the complexity and the order of nature that's obvious in any glance at living things. It's all just too complex to come from chance, from random evolution.

But we know that a glance at a flower or moose or meadow isn't enough to appreciate all of nature, just as a glance at a book isn't enough to appreciate a whole story. A glance at a living thing sees the here and now, but is blind to the billions of years of life recorded in the fossil record, or even the circumstances of life outside our immediate surroundings. Writhing, splitting, and fusing DNA; clouds of fish eggs billowing in the sea; swarms of microbes in a single drop of water—they're all invisible from a simple glimpse of our immediate surroundings, but only the foolish would ignore what they have to tell us about the phenomenon of life. Science allows us to appreciate these worlds. To really understand nature, we have to do more than glance, and we have to think beyond our knee-jerk reaction that insists *since we make things with intent, nature must also be made with intent*. If we want to understand better than a child, we have to look harder, and think deeper. We can see that silt and sand, carried by streams of melting snow, piles up as orderly cones against mountainsides, and we can see that freezing ice crystals make geometrical snowflakes. We have to keep our minds open to the possibility that undirected, random processes can also generate order, even in the domain of living things.

Both supporters and critics of evolution use the same phrase—"evolution is random"—to support their claims. To really understand the phrase we need to distinguish between how it's used to support these opposing viewpoints.

## Randomness in Critiques of Evolution

When we ask how the complexities of life can be accounted for by "random evolution," we're asking a question that itself is fundamentally flawed. That's because the question proposes that evolution is a single force, a unified engine that punches out finished products: beetles, fish, acorns, daffodils. It's natural to wonder how random forces could produce such complex things as beetles, or entire ecosystems.

But the question itself is wrong because it doesn't appreciate the subtleties involved in using the word *evolution*.

As we saw in "Myth Two: It's Just a Theory," an accurate use of the word *evolution* recognizes the cumulative result of three factual processes: the *replication* of life-forms, their *variation* from their parent generation, and the *selection* for or against certain variations, depending on their environment. Obviously, then, the simple phrase "evolution is random" is dangerously clumsy: it simply doesn't acknowledge the complexity of evolution.

The entire question, then, of how "random evolution" can produce something as complex as a mosquito or a lung is something like asking *how could a random mixture of muscle, nerve, skin, and connective cells make a human body?* Without looking carefully at the recipe book (DNA) that tells *how* these cell types are to be ordered to build that body, and without realizing that that body is the product of a long process, rather than an instant in time, we certainly will conclude that random forces couldn't just sweep up these ingredients and make a human being. But for anything more than the most superficial understanding of the human body, we have to understand the form of that body, and the instructions for its construction. This will give us a far better appreciation of the complex system than just a glance at its finished form.

To find the significance of randomness in evolution, then, we have to examine evolution very carefully; it takes time and thought. We have to conceive of evolution not as a whole, a unified endeavor, but as the result of the three fundamental processes of *replication*, *variation*, and *selection*. If there's order or planning, or randomness in the cumulative thing we call evolution, we should find it in these essential elements of the evolutionary process.

## Randomness in Evolution

### Is Replication Random?

We know by now that evolutionary replication is reproduction, that is, the phenomenon of parent generations having offspring. This is a fact that we can see every moment, everywhere on Earth, and randomness or order really aren't relevant here;[3] replication is simply a fact. Parents have offspring, be they lilies or oysters.

As we saw in "Myth Two: It's Just a Theory," mate choice (at least in sexually reproducing species) is an important aspect of replication, as is the question of which individuals of a population actually end up having offspring. But for the moment it's enough to note that randomness just doesn't apply to replication. Replication is like gravity: it simply is; it simply happens. We can move on to look for randomness or for order in *variation*, and then in *selection*.

### Is Variation Random?

We know that when replication occurs, when offspring are produced, the offspring carry DNA very similar to that of their parents. But those offspring are almost never *identical* to their parents.[4] That's because DNA, the molecule that directs the construction of the proteins that build bodies, is changeable.[5] The question is, how do those changes come about? If there's an ordering mechanism, a plan, directing how offspring should differ from their parents, we should see it in the sources of variation, in the sources of change in the DNA that make offspring.

Since variation really refers to the differences between the DNA of parent and of the offspring, we must ask just how DNA actually changes. Although a DNA molecule is very good at making high-fidelity copies of itself,[6] biology has found that there are many ways that it can be changed.[7]

One of the major sources of DNA change is *mutation*, generally defined as any variation in the DNA that's not a result of recombination (the mixing of male and female DNA).[8] Mutations can arise in many ways. There are insertion mutations (where bits of new information are slipped into a sequence) and deletion mutations (where information is skipped). Both change the DNA, and thus the offspring differ from the parents. The environment itself can also cause DNA mutations by the action of certain chemicals, for example, or ultraviolet rays. Viruses, such as those causing chicken pox, can also alter DNA.

Looking closely at how DNA changes, then, we're finding that many factors can alter DNA, making offspring different from their parents. But it's hard to imagine how any plan or consciousness could marshal such diverse forces as ultraviolet rays, blooms, and dissipations of certain chemicals, or the spread and later decline of certain strains of viruses in order to achieve a particular goal.[9] No, what we see here, from the point of view of DNA itself, is a whirlwind of factors that may or may not cause it to change. There's little order, and there's no predicting when a ray of radiation or a viral infection will alter one's DNA.

There's no order here, no intent or designer, but let's keep going. Up ahead there's an entirely different mechanism that causes parent and offspring DNA to differ, and it is profoundly important. Maybe there we'll find order, or a plan, that has worked toward building complex forms and even entire ecosystems.

This other major source of variation is *recombination*. It is the mixing of male and female DNA in sexually reproducing species, creating a unique variation on the mother/father theme.[10] When the female egg cell (the ovum) admits a single sperm cell, the ovum seals up, shutting out any other suitors. Then the male DNA carried in the sperm cell and the female DNA residing in the egg cell not only fuse, but "shuffle." If the mating is successful, a new individual, a child, takes form. It is unique, carrying the "shuffled" DNA of its parents. With this kind of "reshuffling" of the genetic deck of cards at each generation, it's no wonder parents and offspring differ.

The question is, do we see a plan here, an intent or a guide that will dictate how the DNA will be "shuffled" and what mutations might arise? No. Exactly what sort of differences emerge in the offspring is eminently random. That's because neither mutation nor recombination can see into the future and make plans.

Again, the search for intended design comes up empty, here in the realm of recombination. We're narrowing down the possibilities, and we've arrived at the third major process that we subsume under the single word *evolution*. Since we see so much order in the world of life, its source must be here.

## Is Selection Random?

The birth of an individual—an acorn, a puppy, or a jellyfish—is the instant immersion of that individual into a staggeringly complex environment of shifting selective pressures. We saw in "Myth One: Survival of the Fittest" that *fitness* could be measured as an individual's probability of passing on its genes to the next generation, and that its fitness score changes with even the subtlest changes in the selective environment. We then left the concepts of *selective environments*, *fitness*, and *selection*, but it's time to revisit them, as they're central to understanding how the evolutionary process actually does create order, and even progress.

When a globule of frog eggs hatch, and tadpoles begin to explore their surroundings, vigorously flitting around to find and feast on algae or other small organisms, we know from all kinds of observations that those pollywogs aren't clones, even though we might need a magnifying glass, or even a microscope, to see their variations on basic *pollywogdom*. Not only does each tadpole differ slightly from its parents, but also from its brothers and sisters. Due to any number of random factors, such as mutation or recombination, some pollywogs will have slightly better eyesight, or later in life slightly shorter tongues or longer legs than their siblings, and so each has a slightly different chance of passing the genes that made them on to the next generation. Obviously, not just any mutation can crop up; the evolutionary heritage of the species has shaped the DNA to produce basic pollywog shapes, and a pair of wings just aren't going to sprout up one day, as pollywogdom just doesn't have any genetic proclivity or preconditioning, as it were, toward flight. But over long periods of time, new body forms can indeed result. The most important thing to remember here is that those variations arise randomly, rather than according to any perceptible plan.

But while variations on pollywogdom do arise randomly, *which tadpoles survive will not be random*. This is because the selective environment has certain conditions that must be met: tongues must be so long to snatch up prey, legs must be so muscular to make snake-escaping leaps, and so on. Selection doesn't randomly determine which pollywogs will survive and which won't. Rather, it simply evaluates the fitness of each individual, allowing some to reproduce and barring others. Selection selects *for* the better suited and *against* the less well suited, and its effects are certainly not random. Selection preserves the genes that built the bodies better suited for the environment.

The implications are breathtaking and profound. Here, then, is nothing less than the engine of order in life and living systems. Natural selection is the sorting of living things, the evaluation of the phenotypes (bodies), and, of course, of the genes that made them. The results of selection are not random; selection culls the less-fit phenotypes, sparing more fit phenotypes that may survive and produce the next generation.

So, at first, our search for an ordering agent in the evolutionary process was coming up dry. Randomness didn't really apply to the fact of replication and we saw that so many things could change DNA that intent or planning in its variation was literally unbelievable. But when we came to selection, we found that the effects of random selective agents (like a meteorite from outer space or a viral change to DNA) on randomly varying populations of offspring were not random at all. Those who were best suited to an environment survived and passed on their genes. Evolution, in this sense, isn't random. Quite unconsciously, without any intent, the forces of evolution shape randomly varying populations into enormously complex and diverse ecosystems.

## Is Evolution Random?

We began this chapter with the common question *how can random evolution make complex things?* But we quickly found that this question just doesn't cut it; it just doesn't recognize what's really involved in evolution. The question was clumsy and essentially unanswerable. A better series of questions is: *Is replication random? Is variation random?* and *Is selection random?* And we found that the answers were, essentially, *irrelevant*, *yes*, and a resounding *no*, respectively.

But the evolutionist's claim that "evolution is random and undirected," while probably just a sloppy use of the phrase, remains confusing. Technically, it really means that the sources and the results of mutation and recombination are random and not directed in any particular direction. But selection, we've seen, is not random. Further, as plant and animal life-forms interact, they form complex relationships and entire ecosystems, ecosystems that, as we will see in "Myth Seven: Nature's Perfect Balance," also possess so much complexity and order that it's almost impossible to resist the idea that they are designed.

But if we want to really understand our universe, we have to resist that illusion. Resisting it isn't entirely modern. Around 80 BCE, the Roman philosopher Lucretius noted that

> there is one illusion that you must do your level best to escape . . . you must not imagine that the bright orbs of our eyes were created purposely, so that we might be able to look before us; that our need to stride ahead determined our equipment with the pliant props of thigh and ankle. . . . In fact nothing in our bodies was born in order that we might be able to use it, but the thing born creates the use.[11]

He was right, nearly two thousand years ago, and today we have the scientific evidence to prove it.

The phrase "evolution is random" is poor shorthand for what really happens in nature. The full diversity of life, its order and equilibrium and complexity, is the result of nonrandom selection of random variations in living things.

## Notes

1. K. R. Miller and J. Levine, *Biology*, 3rd ed. (New Jersey: Prentice-Hall, 1995), p. 658.

2. See P. Frank, *Einstein, His Life and Times* (New York: Knopf, 1947), p. 209.

3. Although many individuals can, and do, carefully pick their mates (this phenomenon, called *sexual selection*, is discussed in "Myth Three: The Ladder of Progress"), in many species this isn't the case, and mate choice is essentially random. Since fitness, for most species, is a concept that works in the present and in the past (because only humans can anticipate the future), mate selection is made for characteristics *that worked in the previous or works in the present environment* only.

4. In strictly asexually reproducing species, which don't mix male and female DNA (because there are no males or females), many offspring *are* essentially clones, but this doesn't mean that variation never occurs. In sexually reproducing species, identical twins are very rare; they result not from a lack of new combinations of DNA from the male and female parents, but from the formation of two individuals from a *single* fertilized egg.

5. "Bodies," here, can be anything from leaves to single-celled amoebas or multicellular animals.

6. The extraordinary fidelity of DNA replication is discussed in R. Dawkins, *The Selfish Gene* (New York: Oxford University Press, 1976). For a technical summary showing that error rates for DNA can vary from one mutation per thousand generations to one per million generations, see T. A. Kunkel and K. Bebenek, "DNA Replication Fidelity," *Annual Review of Biochemistry* 69 (2000): 497–529.

7. Here it is useful to introduce the "central dogma of biology": that

*genotype* generates *phenotype*, which is evaluated by a selective environment. In other words, the DNA (genotype) directs the construction of a body (phenotype), and that body is selected for, or against, by a selective environment.

8. The word *mutation* has a negative connotation in common use, but in biology it simply means a novelty. Mutations can provide useful new characteristics, or neutral ones, or damaging and lethal characteristics. Our summary of mutation is based on M. W. Strickberger, *Genetics*, 3rd ed. (New York: Macmillan, 1985), pp. 457–505.

9. For information on ultraviolet radiation damage to plants, see A. E. Stapleton, "Ultraviolet Radiation and Plants: Burning Questions," *Plant Cell* 4 (1992): 1353–58. For chemical mutagens in plants, see A. Britt, "DNA Damage and Repair in Plants," *Annual Review of Plant Physiology and Plant Molecular Biology* 47 (1996): 75–100, and for viral infections in plants, see L. Bos, ed., *Plant Viruses: Unique and Intriguing Pathogens: A Textbook of Plant Virology* (Leiden, Netherlands: Backhuys, 1999).

10. Note that members of some asexually reproducing species do occasionally transfer genetic material between individuals. When certain individuals of the bacteria *Escheria coli* (which thrive in the human intestine) come into contact with one another, genetic material is passed from the *donor* to the *recipient*. The recipient may then display characteristics of the donor, and the recipient's offspring will also carry these modifications. This was first reported in 1946. See J. Lederburg and E. L. Tatum, "Gene Recombination in E. coli," *Nature* 158 (1946): 558. Generally speaking, however, recombination occurs only among sexually reproducing organisms, those with male and female individuals. Sex, it seems, started roughly a billion years ago. See A. H. Knoll, "The Early Evolution of Eukaryotes: A Geological Perspective," *Science* 256, no. 5057 (1992): 622–27. Today there are millions of sexually reproducing species.

11. Lucretius, *On the Nature of the Universe*, trans. and ed. R. E. Latham (London: Penguin Books, 1982), p. 156.

## Myth Six

# People Come from Monkeys

It's the summer of 1860, in Oxford, England, and Darwin's *On the Origin of Species* has been in print for only seven months. Over five hundred people—men in somber suits and women in summer dresses—have assembled to hear the latest scientific findings at a meeting of the British Association. At the podium Bishop Samuel Wilberforce (1805–1873) lays into Darwinism, the daring theory implying that humans are not divine creations, but instead evolving animals. Wilberforce achieves a crescendo when he pointedly asks, was it "through his grandmother or his grandfather that he was descended from a monkey?"[1]

Wilberforce's shrill comments could be expected from clergy in 1860. After all, Darwin's theory challenged some basic concepts that had been long held by Western civilization, and the Christian Church in particular, for nearly two thousand years: that life—including humans—was created by divine act, and that humans were divinely special, distinctly different from the rest of animal life. Even though the church had already backed away from a lot of clearly untenable positions—like the view that Earth is at the center of the universe—it still hadn't managed to let go of "human specialness."

Since 1860 the same basic scientific method that has identified how to make planes fly and computers calculate has thoroughly tested the validity of evolutionary theory, and the overwhelming consensus is that, yes, Darwin was right. With nearly a century and a half of experiments and observations supporting evolution, as well as the connections between humans and other primates, you wouldn't think that Wilberforce's old argument—against any human connection with "monkeys"—could still be dredged up in an attempt to discredit evolution. But in summer of 2004, in Dover, Pennsylvania, a disgruntled public school board member complained that evolution shouldn't be taught in public schools because it teaches that people "come from monkeys and chimpanzees."[2]

Neither in Oxford nor in Dover was any flaw of evolutionary theory itself pointed out. No evidence was produced that separated humans from other primates. What connects Oxford in the summer of 1860 and Dover in the summer of 2004 is that in both times and places, evolution was misrepresented, as well as criticized by a purely emotional reaction to its implications.

To evaluate this myth—that evolution claims "we come from monkeys"—we can start by asking what evolution really says about our connection to the rest of living things. Once our place in evolving nature is revealed, we can examine the second claim, that being related to monkeys is a degradation.

## Our Place in Nature

The theory and study of evolution have cast light on a thousand fascinating facts that have helped flesh out our understanding of the natural world. For example, we know that viruses survive by hijacking the DNA of other organisms, that Antarctic penguins evolve very quickly, and that sea slug populations rocket and crash depending on the interactions of viruses and other small organisms that thrive inside them.[3] Each of these facts is astonishing and fascinating in its own right.

But one thing that science has discovered is so completely astonishing, especially to those who still see humans as special creations, that many widely reject it today. It is the fact that *humans are animals*. We often draw a conceptual line between ourselves and other living things, and the word *animal* is usually used to mean something decidedly nonhuman. As expressions of scorn, we talk of *animal nature*, or we say *those people behaved like animals*. But science and evolution tell us that the line between us and the rest of the animals is dotted, at best. It's not solid. Whether we like it or not, we're animals, something even Aristotle recognized over 2,200 years ago. We *have to be*: we're not minerals, as the saying goes, nor are we vegetables. Unless we believe we're special creations that popped up once and have lived on unchanged ever since—which the theory of evolution counters in principle, and which the fossil record refutes—we must be animals. And unless we want to ignore museum vaults full of geological and fossil discoveries showing that species evolve through time, then we must admit that we, too, have an evolutionary past. How can we find out about our past and our position in the present world of living things? We can begin simply by surveying the world of living things and attempting to locate our position within it. Of the living things, what are we humans least like, and what are we most like?

Before we begin, we have to remember to be careful with how we interpret what we find. We humans have long recognized the differences between ourselves and every other kind of life, and what we've thought of those other kinds hasn't always been complimentary. In their book on human origins, *Shadows of Forgotten Ancestors*, Carl Sagan and Ann Druyan describe the Victorian attitude toward chimpanzees by quoting Boston physician Thomas N. Savage, who characterized chimpanzees as "vile," "degenerate," "filthy," and "depraved."[4] Throughout the history of Western civilization, we have looked suspiciously on nonhuman primates, sometimes envisioning them as devils, sometimes as misshapen, half-human wretches created by God as a warning against sin.[5] Even when we look at animals other than primates, we have an obvious hierarchy of preferences; we love mammals, particularly those with big eyes and soft fur a bit more than we love, say, fish, but we love fish more than ants or spiders.[6] In the past, as we groped for a meaning or explanation of apes and of all other living things, we kept them at a distance. But if we're going to take an honest look at the rest of life, and ourselves, and our connections with the rest of life, we have to leave these kinds of emotional evaluations behind. We have to have the courage to look not only into the mirror, but also through our family history, through trunks of ancient letters and newspaper clippings. And we must think about what we find in there.

Where to start? Everywhere we look, we see life. Stopping life from multiplying and overrunning us seems to be a constant task; we mow our lawns, weed our gardens, brush our teeth, and wash our dishes, all to keep down one kind of life or another. How do we find our place in the natural world, the world that teems with millions of living species?[7]

Clearly, we have to start by looking at obvious groupings; for example, things that we can eat and things that we can't eat. But there are so many types of life out there, from slime molds to pine trees to hawks. And, of course, we could classify things in any number of ways: by their size, for example, or where they're found, or even how they taste (sweet, sour, or tart)—like the English eccentric William Buckland, who attempted to eat his way ". . . right through the animal kingdom."[8] But are all of these ways of classifying equally valid? Obviously not. A lot of classifications would be transparently self-serving, and if we're intending to find our place in nature rather than assuming we're at its center, or pinnacle, then we need an objective way of grouping life—a way that doesn't judge species by their usefulness to us.

This was exactly what Carolus Linnaeus (1707–1778), a Swedish medical student, accomplished in 1735, when he published *Systema Naturae*, his classification of all known living things. The Linnean system was revolutionary. Like Darwin's later challenge to human specialness, it was a challenge to another ancient belief, the *Great Chain of Being*, which (as we saw in "Myth Three: The Ladder of Progress") classified living things by the degree of perfection they were thought to possess. But Linneaus classified things by how they looked and by their outward characteristics,[9] not their degree of perfection, and he did it because the Great Chain of Being simply couldn't stretch to account for new discoveries, such as species that weren't mentioned in the Bible.

Today we continually update the Linnean system with the discoveries of new species and the clarifications of relationships based on DNA studies.[10] But we haven't abandoned the basic approach of organizing living things by their physical characteristics. Let's see what that approach tells us about our relationship to the rest of the living things on Earth.

## A Journey Through the Animal Kingdom

We can take a fantastical journey, now, to visit the constellation of life-forms on Earth. We can travel high into the air, squirm through the vines and roots of dripping jungles, or plunge down to the seafloor, and everywhere we go, we'll find life. In the real world, we'd need special equipment: scuba gear, ventilators for gas-choked caverns, insulating clothes to protect us from the Arctic cold. But in the mind, we can move freely. What do we find?

At the start of our journey through the Animal Kingdom, we stand at the first fork in the road. One sign points left, declaring: "Subkingdom Protozoa." The other points right: "Subkingdom Metazoa."

If we went left, we'd plunge into the world of unicellular microbes, bacteria that swarm and multiply in even the most unlikely places; we find them wherever we lift a rock, crack apart a piece of ice, spoon up a gob of boiling mud, or swipe through the air with a jar.[11] These are life's Xerox copies, organisms that bud off near replicas without combining their genetic material with a mate. They are innumerable and diverse. We know there are over fifty thousand species of bacteria.[12] They spawn in the air we breathe and within our own bodies. They can reproduce in a matter of hours, whereas we humans have to wait over a decade before reaching sexual maturity. And they're tiny, invisible to all humans until the invention of the microscope in the late seventeenth century.[13] We commonly know them for the harm they do to us: some are responsible for diarrhea, some for pneumonia. But others are beneficial to us; they break down nutrients in soil, making agriculture possible, and others aid in our digestion.

We don't follow the sign on the left, though; we don't belong in that world of anonymous blobs. We're composed of billions of cells, not just one, and we reproduce sexually, not asexually. So we step right, making our way into the *Subkingdom Metazoa*.

In this realm we find more familiar sights: creatures made of dozens or thousands or even billions of cells. Most are highly mobile, some able to pursue prey, others able to escape predation. It's an enormous realm, crowded with every multicellular mobile creature that has ever lived, or that lives today,[14] from whiplike nematode worms (they live in your backyard, at the bottom of the Mediterranean sea, and everywhere in between) to clicking lobsters, hairy spiders, and warm, furry guinea pigs. We search for familiarity in the chaos, and we find thirty-five major groupings, each called a *phylum*—things we're like, and things we're definitely not like.

We're not much like anything in the Arthropod phylum, but that just makes us the minority. A million arthropod species have been named, and it's been estimated that there could be nine million species of them in tropical forests alone,[15] and perhaps thirty million species worldwide. These multitudes of spiders, insects, crabs, and lobsters scuttle and skitter on jointed appendages, and, compared to us, they're very strange. They're inside out! Think of a scorpion, with its skeleton on the outside, protecting its soft tissues on the inside. No, we don't belong here.

We keep looking. There's the Mulluscan phylum, which includes ingenious octopi, hardy oysters, and slow-creeping garden snails. But the octopus's fluidity and the seashells that many mollusks inhabit have strikingly different characteristics from our four-limbed bodies; again we have to keep searching for something familiar. We find other phyla, one containing corals and jellyfish, but the coral's immobility and the jellyfish's watery, baglike form are equally alien. We don't find much in common in the Bryozoan phylum either, a group of about five thousand species of mosslike animals that live in colonies beneath the sea.

But now we come to something that does look familiar, if only vaguely: creatures with a nerve cord running down the middle of the body, a sort of spine. Although not all of the members of this phylum —the *Chordata*—have skeletons, they all do have a nerve cord, and humans share it and a few other characteristics with them. Of the thirty-five forks in the road, we take the path marked "Chordata," and immediately we come to another fork: "Invertebrate Subphylum," says one sign; "Vertebrate Subphylum," says the other.

Peering down the road of the Invertebrate subphylum, we see almost thirteen hundred species, from the sea squirts (inch-diameter "sea puffballs" that filter water for food particles) to the eel-like cephalochordates, first identified from fossils over five hundred million years old,[16] and today harvested and cooked as food in South Asia. We're more like any of these than we are like the jellyfish, worms, or bacteria we glanced at earlier, but they're still strange to us, and very different from what we see ahead on the road leading to the Vertebrates. There, we vaguely perceive another great swarm of life, millions of species, but all with one great similarity to ourselves—they have distinct spines. We follow the sign "Vertebrate Subphylum."

Here are the things we easily recognize as animals. There are seven main classes. Nearly four thousand species of amphibians hop and squirm in waterways and on muddy shores. They include species as surprising as the wood frog, which can survive being nearly frozen solid for weeks at a time.[17] But we've been out of the water and the mud for a long time, and so we keep looking for closer relatives. Other classes contain the bony fishes (like tasty salmon), and cartilaginous fishes (like sharks); the reptiles (everything from dinosaurs to tortoises, snakes, and crocodiles: they are the producers of hard-shelled eggs); the birds (who also produce hard-shelled eggs, but are mostly adapted to flying as their means of getting around); and the mammals, vertebrates with hair, self-warming bodies, mammary glands for feeding the young, and relatively large brains. The mammals—now we're getting close to home. Of the seven groups of vertebrates, of the seven forks of the road, we confidently stride down the one leading to the "Mammalian Class."

Soon we arrive at three more forks, but it's getting easier to find our way now. We see three main subclasses of the mammals: the Prototheria (including the platypus and the spiny anteater), the Metatheria (the marsupials, such as the kangaroo and the opossum), and the Eutheria (the mammals whose young spend a long time developing in the placenta). We go straight down the road of the Eutheria.

We find more wonders here, but they're less-alien wonders.

There are at least nineteen orders of eutherian mammals.[18] Among them are the Cetacea, the sea mammals in whom we recognize something familiar, but more immediate to us are the Perissodactyla (*periss*, which in Greek means "odd," and *dactyl*, which means "toed"), which are the tapirs, rhinos, and horses. We know zebras almost instinctually, having lived with them on the savannas of Africa for millions of years, but it's the sight of horses that touches us emotionally. We first domesticated them around six thousand years ago,[19] and since then they have accompanied us in countless triumphs and disasters. And there are the Artiodactyls (even-toed), including our beloved cows and pigs and handy camels, and another familiar group containing friend and foe alike, the Carnivora, characterized by teeth adapted for slashing meat. This group also contains animals that touch us emotionally: bears and dogs (first domesticated over fifteen thousand years ago),[20] and cats (certainly domesticated by the time of Egyptian civilization, over three thousand years ago, and maybe as early as nine thousand years ago).[21] It seems like we're home now among the animals that we know so well. But a moment of thought reminds us that we haven't taken this journey to find the animals we're comfortable with; it's a journey of self-discovery, an attempt to find what we're most like, and least like. So we have to keep looking down the roadways of the nineteen mammalian orders.

And then we find it, the group containing the mammals we most resemble, the monkeys and apes. Of the nineteen roads, we take the fork signposted "Primates."

Here we find the vertebrates that are eutherian mammals sharing an array of traits, including dexterous hands, nails rather than claws, and reliance on vision rather than sense of smell.[22] We're like these creatures, and we're close to home now.

But again we come to another fork almost immediately, the left fork posted "Suborder Prosimii," the right, "Suborder Anthropoidea." Looking down the prosimian road, we see lemurs—catlike primates stranded on the islands off East Africa—and lorises and galagoes (the *Star Wars* Ewoks must have been modeled on these "bush-babies"), as well as the strange tarsiers. These prosimians resemble the earliest primates of all, known to exist over sixty million years ago, and while they're clearly primates, they're also very different from us. We take the right fork, down the road of the "Suborder Anthropoidea."

Of the 233 known and living species of primates, 145 (62 percent) are anthropoid (*anthropo* meaning "human," *oid* meaning "like"). These are the monkeys, the apes, and the species they most resemble —humans. We can almost jog now, easily picking the forks in the road. The fossil record and the DNA evidence make it clear that humans didn't originate in the Americas, so at the fork dividing the New and Old World primates, we head for the Old World, the world of the Catarrhinni (named by Linnaeus for their "downward-facing nostrils"), passing by more than thirty species of South American primates, many of which are tree-dwelling monkeys we call howlers, or spiders, depending on their lively vocalizations, or their use of a tail like a fifth hand.[23]

We hardly have to slow down as we find the next major fork in the road, one leading to the monkeys, the other to the apes. Of all the primates, we're clearly from the Old World, and of those from the Old World we're more like the apes (the Hominoidea) than the monkeys (the Cercopithecoidea). Although the monkeys are very familiar, and we can read emotion in their faces, we differ from them in important ways. Unlike baboons, for example, we don't have enormous canines or tails. Also, we're much larger than any monkeys, and of course, we walk upright.

So we jog right past the monkeys and down the road leading to the Hominoidea (the "humanlike" primates). But here we have to slow down. The species are starting to look similar to one another, and suddenly it's harder to tell which we're most like or unlike. Only one group is easy to rule out. These are the gibbons (found in South Asia), whose most striking difference from us is that they're brachiators, cruising like Tarzan through the treetops, swinging from branch to branch. Although many of us like to climb things, we don't live in the trees, and we don't move by brachiation. We can also bypass the highly arboreal (tree-dwelling) orangutan of Southeast Asian islands. Although they're aboriginally known as the "Man [*orang*] of the Forest [*utan*]" for their similarity to people, they're tree-dwellers, while we're firmly committed to the ground. We move on once again.

Now we come to the terrestrial Hominoids, who first appeared in the jungle regions of Africa. Looking back at us, from their green and leafy habitats, are the gorillas and the chimpanzees. We know them well. They're more like us—in anatomy, genetics, and behavior—than any other living thing. And there, crouched behind a bush and also observing the chimps and the gorillas, is a creature even more similar to ourselves. The creature wears clothes and holds binoculars. Wanting a better view, it stands, and we know we've arrived—a human stands before us. We've found our place in the world of the living things, and in the world of mammals, primates, and hominoids.

## Our Primate Ancestors

Our tour of the Animal Kingdom brings us right back to the original contention, often used by antievolutionists, that evolution states that "we come from monkeys." Given what we've seen in the Animal Kingdom and the Primate order, is this accurate?

We're certainly similar to living and fossilized monkeys, but we're much more similar to apes. We can look at the evolution of the Primate order to see why.

An overview of primate evolution, pieced together from the fossil record (and, more recently, DNA studies) sketches out a sixty-five-million-year history of the Primate order, divided into five main "adaptive radiations," periods of substantial change in biological lineages, often driven by environmental changes.[24]

The first primate adaptive radiation occurred shortly after the fall of the dinosaurs, around sixty-five million years ago, when the earliest primates were just diverging from their early mammal ancestors. The primate pattern was being established, and at this time primates were small, insect-eating creatures similar to squirrels. The most important difference from their nonprimate ancestors is seen in their teeth, which show adaptations for processing new food sources. Seeds, fruits, and other vegetation, commonly found in trees and bushes, were added to the insect diet, and the Primate order was under way.

Roughly ten million years later we see the second radiation. Primate faces have changed, with snouts getting smaller and eyes moving closer together, toward the front of the face. Here we have the origins of binocular vision, and a premium on vision over the sense of smell; we also see the origins of extremely dexterous hands and the reduction of claws into fingernails and toenails.[25] There's a divergence here, as some primates remained rather insectivorous and squirrel-like (the ancestors of modern lemurs), while others continued to adapt to more varied diets (the ancestors of all other primates).

About forty million years ago the fossil record indicates the next adaptive radiation, in which we see the origins of larger brains (though they're still about thirty cubic centimeters: about a tenth of the volume of a soda can), and yet further reduction in the sense of smell, but a still greater premium put on vision. And there's a major geographical event here, as well, with the establishment of a new population of primates in the New World, specifically on the continent of South America.[26]

Yet another ten million years pass before we see the fourth significant primate radiation, this one taking place in Africa, around thirty million years ago. Here we see a major divergence in which some primates focus on leaves as their main diet, while others focus on fruits. This is the earliest differentiation between monkeys (the leaf-eaters) and apes, which were somewhat larger, and, at least at first, spent more time in the trees while monkeys came to the ground.

Then, between five and ten million years ago, we see another major radiation, in which environmental changes open new habitats, and the primates radiate again, adapting to new circumstances and environments. Around six million years ago we see the origins of our own *hominid* lineage, characterized most strikingly by *bipedalism*, or standing upright.[27] As Earth's climate changed and grasslands began to fragment the tropical forests of Africa, some of the chimpanzee-human ancestor population moved into the newly emerging terrestrial habitats, where bipedalism succeeded as a new way of life for a new variety of primate: the hominid.[28]

By two million years ago, this lineage had itself diverged into three main types of African bipeds. These were the australopithecines, divided into the lightly built variety (the graciles) that had a varied diet and lived in wooded areas, and a vegetarian variety (the robusts) that spent more time in the open savannah, eating grasses, seeds, and roots.[29] There were also the earliest members of our genus (*Homo*), characterized by large brains, far more complex tool use, ever-smaller teeth, and ever-larger bodies in general. It's clear what happened to these three types of hominids. The robusts went extinct, possibly having overspecialized on a grassy or seedy diet. The graciles also vanished from the fossil record, but this occurs just before *Homo* is found in substantial numbers. It's widely believed that some variety of gracile australopithecines evolved into early *Homo*, our ancestors.[30] The evolution of the Primate order shows us why we look so similar to the other primates, including monkeys—because we share ancestors with them.

## Do People Come from Monkeys?

The simplest investigation of the characteristics of living things makes it clear that we're mammals, and that of all the mammals, we're primates. So we can dismiss some critics' lamentation that believing in evolution will drag us into the world of the primates: we're already in it. And stacks of data make it equally clear, to any thinking person, that people *don't come from monkeys*. Around thirty million years ago, the African primates diverged into two distinct groups, taking up different diets, habits, and habitats, as their environments changed and opportunities arose.[31] One group developed into apes, which included—much later—chimpanzees, gorillas, and humans. The other group developed into monkeys, and DNA and other studies prove that we belong in the ape group instead of the monkey group. Do people come from monkeys? Not at all. We do share a common ancestor with chimpanzees, and before them, with the group that became monkeys. But to say we come from monkeys is simply wrong, and evolution has never claimed it.

Early in this chapter, we saw that Samuel Wilberforce's argument against evolution, echoed again even in 2004, was obviously emotional rather than factual. Unless we want to live in a web of lies, we can't pick and choose what to believe, not when the raw data of genetics, fossil studies, and anatomical studies are laid before us.

Anyone is free to argue whether they *like* being primates or whether they *like* being related to chimpanzees—but that's not the question. The question is whether or not we're descended from monkeys, and the evidence is in: we're not, but we are related to them.

## Notes

1. The quote is from I. Sidgwick, "A Grandmother's Tales," *Macmillan's Magazine* 78 (1898): 433–34, as referenced in J. R. Lucas, "Wilberforce and Huxley: A Legendary Encounter," *Historical Journal* 22, no. 2 (1979). A fuller account of Wilberforce's diatribe is found in W. Irvine, *Apes, Angels and Victorians* (New York: McGraw-Hill, 1955). J. R. Lucas notes that Wilberforce's speech has been embellished and exaggerated (Lucas, "Wilberforce and Huxley"), but it's clear that Wilberforce tried to slight evolutionists by implying that those who supported evolution would drag humanity into the world of monkeys.

2. See A. Badkhen, "Anti-Evolution Teachings Gain Foothold in U.S. Schools," *San Francisco Chronicle*, November 30, 2004, p. A-1.

3. For viruses, see A. G. Fettner, *Viruses: Agents of Change* (New York: McGraw-Hill, 1990); for penguin evolution, see D. M. Lambert et al., "Rates of Evolution in Ancient DNA from Adélie Penguins," *Science* 295, no 5563 (2002): 2270–73; for sea slugs, see S. K. Pierce et al., "Annual Viral

Expression in a Sea Slug Population: Life Cycle Control and Symbiotic Chloroplast Maintenance," *Biological Bulletin* 197, no. 1 (1999): 1–6.

4. See C. Sagan and A. Druyan, *Shadows of Forgotten Ancestors: A Search for Who We Are* (New York: Ballantine Books, 1992), p. 270. In the same book the authors note that the Judeo-Christian-Islamic religious system arose in a region where nonhuman primates were rare or absent.

5. A winged ape-Devil carved from stone in eleventh-century Spain and remarkably resembling the terrifying winged apes in the 1939 film *The Wizard of Oz* is seen in H. W. Janson, *Apes and Ape Lore in the Middle Ages and the Renaissance* (London: Warburg Institute, 1952), pl. IIb.

6. Our attraction to animal cuteness is so strong that it's even exploited by advertisers: consider how many (and what kinds) of animals are used to market breakfast cereals and other products that have nothing to do with those animals in the first place. See a review in G. Feldhamer et al., "Charismatic Mammalian Megafauna: Public Empathy and Marketing Strategy," *Journal of Popular Culture* 36, no. 1 (2002): 160–67.

7. Linnaeus named nine thousand species of plants and animals; today we know of about five million species, but some have suggested this estimate may be ten times too low: see R. M. May, "How Many Species?" *Philosophical Transactions of the Royal Society of London B* 330 (1990): 293–304. We discuss the concept and definition of species in "Myth Four: The Missing Link."

8. For more on Buckland, see N. A. Rupke, *The Great Chain of History: William Buckland and the English School of Geology (1814–1849)* (New York: Oxford University Press, 1983).

9. While Linnaeus classified mostly by outward form, later evolutionists began to include other characteristics into species definitions, and by 1996 Mayr stated that "[d]egree of morphological difference is not an appropriate species definition. . . ." See E. Mayr, "What Is a Species and What Is Not?" *Philosophy of Science* 63 (1996): 262–77. Still, genetic and other modern studies show that Linnaeus correctly identified many species.

10. Since the 1930s, on average, a new mammal species has been discovered about every three years: see R. H. Pine, "New Mammals Not So Seldom," *Nature* 368 (1994): 593. For example, on Tuesday, December 6, 2005, the BBC online news reported the discovery of a new mammal species in Borneo's Kayan Mentarang National Park: see R. Black, "'New Mammal' Seen in Borneo Woods," http://news.bbc.co.uk/1/hi/sci/tech/4501152.stm

(accessed December 6, 2005). It was long assumed that New Zealand's moa and kiwi birds were very closely related, because of their appearance and habitat, but new DNA evidence suggests that kiwis are more closely related to some African birds, and that they and moas arrived in New Zealand many millions of years apart: see A. Cooper et al., "Independent Origins of New Zealand Moas and Kiwis," *Proceedings of the National Academy of Sciences* USA 89, no. 18 (1992): 8741–44.

11. The conditions in which we find life can be astounding: on the frigid, lightless, intensely high-pressured seafloor, we find life flourishing among "extremes in temperature, hypoxia [lack of oxygen], sulfide, and heavy metals." See E. R. McMullin et al., "Metazoans in Extreme Environments: Adaptations of Hydrothermal Vent and Hydrocarbon Seep Fauna," *Gravitational and Space Biology Bulletin* 13, no. 2 (2000): 13–24. For an introduction to life's most extreme adaptations, see D. A. Wharton, *Life at the Limits: Organisms in Extreme Environments* (Cambridge: Cambridge University Press, 2002).

12. One biologist has estimated that an ounce of forest soil contains over five hundred thousand species of bacteria, and that worldwide there are probably a billion species! See D. E. Dykhuizen, "Santa Rosalia Revisited: Why Are There So Many Species of Bacteria?" *Antonie van Leeuwenhoek International Journal of General and Molecular Microbiology* 73 (1998): 25–33. The actual count of species of bacteria is less important, according to Ward, than the fact that the count is very large. See B. B. Ward, "How Many Species of Prokaryotes Are There?" *Proceedings of the National Academy of Sciences USA* 99, no. 16 (1999): 10234–36.

13. A wide range of magnifying devices was invented around the late seventeenth century; Dutchman Antonie van Leeuwenhoek is generally credited with devising the first functional microscopes, opening the doors to an astounding world of *animalcules*, his word for microbes. See C. Wilson, *The Invisible World: Early Modern Philosophy and the Invention of the Microscope* (Princeton, NJ: Princeton University Press, 1995).

14. As early as 1954, Simpson estimated that between fifty million and five billion species have existed since the origins of life on Earth: see G. G. Simpson, "How Many Species?" *Evolution* 6, no. 3 (1954): 342. His estimate is probably low.

15. See C. D. Thomas, "Fewer Species," *Nature* 347 (1990): 237.

16. See J. Y. Chen et al., "A Possible Early Cambrian Chordate," *Nature* 377 (1995): 720–22.

17. D. A. Wharton, *Life at the Limits*, pp. 180–85.

18. In an example of how molecular biology is refining our understanding of relationships of living things, a recent attempt to use genetic data to classify the eutherian mammals groups them into four major *clades*: the *Afrotheria* (mostly in Africa), the *Xenartha* (New-World [North and South American] sloths, anteaters, and armadillos), the *Gilres* (rodents and lagomorphs [rabbit-like creatures]), and *Group IV*, which is basically "everything else," from the Cetacea (sea mammals) to the Carnivora, the Chiroptera (bats), and the Insectivora (such as shrews and moles). See W. J. Murphy et al., "Molecular Phylogenetics and the Origins of Placental Mammals," *Nature* 409 (2001): 614–18.

19. Most scientists believe that early horse domestication took place in central Asia. For example, see M. A. Levine, "Botai and the Origins of Horse Domestication," *Journal of Anthropological Archaeology* 18, no. 1 (1999): 29–78. Recent DNA studies, however, suggest that although central Asia was one center of horse domestication, horses may have been domesticated in several different places at different times in prehistory: see T. Jansen et al., "Mitochondrial DNA and the Origins of the Domestic Horse," *Proceedings of the National Academy of Sciences* USA 99, no. 16 (2002): 10905–10.

20. It's hard to be sure when dogs were first domesticated, because early domesticated dogs and wild dogs (wolves) would not be very different skeletally. Some suggest that dogs were first domesticated over three hundred thousand years ago, while others suggest it was closer to twenty thousand years ago, but almost nobody argues it was less than fifteen thousand years ago. For an overview, see J. Serpell and P. Barrett, eds., *The Domestic Dog: Its Evolution, Behaviour, and Interactions with People* (Cambridge: Cambridge University Press, 1995).

21. For an overview of the domestication of mammals, see J. Clutton-Brock, *A Natural History of Domesticated Mammals* (Cambridge: Cambridge University Press, 1999). For new evidence for possible cat domestication long before Egyptian civilization, see T. Rothwell, "Evidence for Taming of Cats," *Science* 305, no. 5691 (2004): 1714.

22. A slightly dated but enjoyable overview of the primates is found in M. Kavanaugh, *A Complete Guide to Monkeys, Apes, and Other Primates* (New

York: Viking, 1984). A more technical overview is available in R. Jurmain et al., *Introduction to Physical Anthropology* (Belmont, CA: Wadsworth, 1999): 106–42.

23. New World primates probably arrived in the Americas, from Africa, more than thirty million years ago. They likely rode natural "rafts" of floating vegetation across the South Atlantic (see note 26), or opportunistically spread from one island to another as the two continents drifted apart. By thirty million years ago, however, African and New World primates were totally isolated. See R. Jurmain et al., *Introduction to Physical Anthropology*, pp. 126–27.

24. For a discussion of adaptive radiation, and an example, see D. Schluter, "Ecological Causes of Adaptive Radiation," *American Naturalist* 148 (November 1996): S40–S64. This review of primate evolution is based on R. G. Klein, *The Human Career* (Chicago: University of Chicago Press, 1999).

25. Binocular vision is most developed in the primates, allowing extremely acute perception needed in their leafy, three-dimensional world of branches and limbs. See C. P. Heesy, "On the Relationship between Orbit Orientation and Binocular Visual Field Overlap in Mammals," in T. D. Smith et al., *Evolution of the Special Senses in Primates: The Anatomical Record Part A: Discoveries in Molecular, Cellular, and Evolutionary Biology* 281A, no. 1 (2004): 1104–10. The "Visual Predation Hypothesis" suggests that binocular vision and other characteristics were important adaptations reflecting the primates' focus on eating hard-to-catch insects.

26. It has been suggested that the fundamental differences between New World and Old World monkeys could have occurred in Africa *before* the two populations split. See M. Takai et al., "New Fossil Materials of the Earliest New World Monkey, *Branisella boliviana,* and the Problem of Platyrrhine Origins," *American Journal of Physical Anthropology* 111, no. 2 (2000): 263–81. One recent study suggests that New World primates could have survived floating on rafts of vegetation from the Old World, across the Atlantic to the New World; they would not, of course, have been doing this as a conscious decision. See A. Houle, "The Origin of Platyrrhines: An Evaluation of the Antarctic Scenario and the Floating Island Model," *American Journal of Physical Anthropology* 109, no. 4 (1999): 541–59.

27. How and why bipedalism (walking habitually on two legs) evolved is one of the greatest questions of anthropology. The topic is introduced in

Jurmain et al., *Introduction to Physical Anthropology*. A recent review finds that while we have a good idea of the dating and the sequence of the evolution of bipedalism, different varieties of bipedalism occurred in hominid evolution, and no current explanations are entirely convincing. See W. E. H. Harcourt-Smith and L. C. Aiello, "Fossils, Feet and the Evolution of Human Bipedal Locomotion," *Journal of Anatomy* 204, no. 5 (2004): 403–16.

28. Evidence from a wide array of fossilized plants and animals clearly shows that by sixteen million years ago, grasslands began to expand in Africa, gradually replacing more forested habitats. By eight million years ago, grasslands were widespread, and many plant and animal species (including ancestors of the early hominids) were adapting to these conditions. For an overview, see B. F. Jacobs, "Palaeobotanical Studies from Tropical Africa: Relevance to the Evolution of Forest, Woodland, and Savannah Biomes," *Philosophical Transactions of the Royal Society of London B* 359 (2004): 1573–83.

29. The earliest discoverers of australopithecine fossils were South Africans, and they named the hominids Southern (*austral*) Apes (*pithecus*). Later, australopithecine fossils were found as far north as Ethiopia. Of the discovery in South Africa in August 1936, paleontologist Robert Broom wrote: "I was again at Sterkfontein, and when I saw Barlow, he handed me a beautiful brain-cast, and said 'Is this what you're after?' I replied, 'Yes, that's what I'm after.' It was clearly the anterior two-thirds of the brain-cast of an anthropoid ape or ape-man, and in perfect condition." See R. Broom, *Finding the Missing Link* (London: Watts, 1950), p. 45. For more on australopithecines, see "Myth Four: The Missing Link."

30. Evidence for the transition from gracile australopithecines to early *Homo* is reviewed in H. M. McHenry and C. Coffing, "*Australopithecus* to *Homo*: Transformations in Body and Mind," *Annual Review of Anthropology* 29 (2000): 125–46.

31. Fossil and genetic data place the divergence of the ancestors of modern Old World monkeys and apes to around thirty million years ago. See M. E. Steiper, N. M. Young, and T. Y. Sukarna, "Genomic Data Support the Hominoid Slowdown and an Early Oligocene Estimate for the Hominoid-Cercopithecoid Divergence," *Proceedings of the National Academy of Sciences USA* 101, no. 49 (2004): 17021–26.

## *Myth Seven*

# Nature's Perfect Balance

For as long as humanity can remember, Earth has been a pretty safe and benign home. We've endured some crises, but we're still here, and the natural world we venerate in everything from our myths to our art is characterized by balance and order. The Nile rises every spring—as it has for millennia—irrigating its fertile banks; the global average temperature hovers around sixty degrees Fahrenheit, which we can offset with a little clothing; salmon spawn and run on a predictable schedule that sustained Native Americans for thousands of years. We marvel at the harmony we see among communities of plants and animals. When undisturbed by humankind, it seems, Earth is at peace, in balance. It's no wonder that we've long believed that nature itself—often personified as "Mother Nature"—

tends to our wants and wishes, carefully managing ecological communities, carefully maintaining *nature's perfect balance.*[1] Since all living things are the product of evolution, many feel that evolution has the purpose of tailoring species—even entire ecosystems—to carry out specific roles in a vast, complex, harmonious, self-regulating system.

But, like waking from a dream, centuries of research have shown us that this conception of purposeful, natural harmony is an illusion, a fabrication of our species' recent and selective memory. Although writing and oral traditions have given us a better and longer memory than any other creatures, our memory still only goes back a few thousand years.[2] Since life on Earth appeared over three billion (that is, three *thousand million*) years ago, our cultural memory samples only about *one thousandth of 1 percent* of the history of life. We've only just started using science to investigate that history, and what we've found is that the past was rife with calamity and extinction, and very different from the harmonious balance we revere today.

If nature was trying to maintain balance, it would mean that evolution has purpose and intent, and that would be very important to know. If this was true, we should see that intent in the unfolding of evolution, both in the way communities of plants and animals coevolve and in the history of any individual species.[3] In short, we should see clear intent in the workings of an immense and complex living thing, known to some as *Gaia*. Luckily, generations of biologists and other observers of nature have accumulated countless reams of information on the natural world as it is today and as it was in the past. We can sift their findings to evaluate the phrase "the balance of nature."

## Where's the Balance? The Fossil Record

Once again the fossil record of ancient life—plant, animal, and the strange things in between—can help us sort out the truth of life on Earth. And we simply must look back in time because looking at the present, or even the last five thousand years of recorded history, only samples a moment of the history of life on Earth. Since most species that have existed have gone extinct, and many of those species were fossilized, we have a great opportunity here—to look for purpose or intent in evolution and nature in the billions of years of life that have already passed. The search for purpose, then, should be a search not just of the living moment, but of the full history of life, written in the fossils. Luckily, we have great data. The fossil record provides us with over two hundred thousand examples of extinction.[4] What do we see in this extraordinary and invaluable encyclopedia of life history? Do we see intent and careful planning? Do we see the workings of an enormous self-regulating system, the evolution of life-forms to carry out certain functions?

The answer is clearly no. Rather than an unfolding plan, what we see is the rise and fall of every imaginable kind of plant and animal—from the giant sloths of South America to the woolly mammoths of the ice age—with their evolutionary origins and demise serving no apparent function or purpose. Some species remain relatively simple through time, with ancient adaptations serving them well for millions of years, while others become increasingly complex. And after about four million years, almost every species goes extinct,[5] and there's little evidence that these extinctions happen according to a systematic plan or schedule.[6] In the fossil record we see species arise and go extinct, apparently without purpose or goal. If there was a goal, we should see it in the *why* of extinctions: it would be relatively clear why this one life-form would be slated for extinction, and that one for survival, as their roles in the system should be clear from their anatomy and—when we could observe it or reconstruct it—their behavior.[7]

When we look at *why* species go extinct, what do we see?

## Where's the Balance? Invasions

Animals don't stay still, and even plant seeds can blow hundreds of miles before taking root.[8] When plants or animals begin to flourish in a new environment, those who might benefit from the new arrivals might consider them friends, but to those who have something to lose, they might be considered weeds, pests, or *invasive species*.[9] When the arriving population can outcompete local populations for the same foods or nutrients, the local population might well go extinct. Is there a pattern, then, or an obvious plan in the arrival of invasive species and their overthrow of native species?

One dramatic example shows that there isn't. For more than a hundred million years, South America was an island continent, slowly grinding northwest through what is today the South Atlantic. In that time it was home to a wide range of marsupials, opossum-like creatures with pouches for their young.[10] Just like mammals on other continents, the marsupials adapted to a variety of habitats. There was even a saber-toothed marsupial cat, *Thylacosmilus*, a predator that fed on other marsupials. But around three million years ago, North American mammals started to arrive in South America, crossing the newly formed land bridge that today we call the Isthmus of Panama. The North American mammalian colonists clearly had some advantages over the marsupials, because soon most of the South American marsupials were extinct. For those South American marsupials, the *balance of nature* completely fell apart, but for the North American mammal immigrants, life just got better and better as they invaded a new continent and replaced the competition.[11]

This is just one case of invasive species penetrating new habitats, wreaking havoc among local populations, driving some locals to extinction, and throwing any balance off-kilter for many generations. And we don't see a plan here, a function or purpose that was served by the evolution of marsupials in South America, and the eventual connection with North America, millions of years later. No rational person could accept such a long and complex plan. So, in this case, as in many others, we see that entire ecosystems were reordered, and not according to a great balancing plan of nature, but simply as a result of geological forces—in this case, tectonic plates sliding around Earth—and of the fact that animals tend to migrate and explore new environments when they appear. There's no plan here.

We humans are familiar with invasive species because we introduce them to new habitats all the time, causing no end of headaches. On June 15, 2005, California State officials were alerted that an alien species had arrived in California. They surrounded a Sacramento County warehouse, evacuated the employees, and blitzed the structure with fumigation bombs. Forest Service personnel fanned out, searching for the invaders in every tree within a quarter mile of the warehouse and placing traps in a nine-mile radius. In the end, a plan was drawn up to monitor the area for three years, to ensure that the aliens had been exterminated. This sounds like a scene from the popular TV series *The X-Files*, but it really happened. What caused this red alert? Slime-spores from deep space? No—it was two Asian long-horned beetles discovered crawling outside the warehouse, which had recently received a shipment of wood from China. According to the Agricultural Plant Health Inspection Service, if Asian Longhorns successfully colonized the United States, they could kill 30 percent of our urban trees, cost timber-related industries billions, and in the end cause "more damage than Dutch elm disease, chestnut blight, and gypsy moths combined."[12]

Like South America crashing into North America, here the appearance of invasive species was a matter of chance, not plans. This is very different from our human habit of consciously importing species to all parts of the globe, in order to carry out specific functions. Farmers know that a ladybug can eat five thousand aphids in its one- or two-year lifespan, so they have introduced them to farms across America. And many of our crops, including sunflowers, mangoes, and apples, are entirely dependent on imported honeybees for their pollination.[13] But this kind of introduction of species into ecosystems, for specific goals, is simply missing from the natural world and the fossil record.

## Where's the Balance? Death from Above

There's another reason that species go extinct, a reason that again has no apparent pattern or purpose. Sometimes death, or even mass extinction, comes from space, from sources so disconnected from Earth that it's hard to argue that they're any part of a natural plan.

At least five major extinctions have occurred in the history of life on Earth, some of them resulting from space debris impact with Earth.[14] The first, around 440 million years ago, wiped out nearly a quarter of the major types of ocean life and seems to have been the result of severe global cooling. The second, around 370 million years ago, might also have been the result of climate change, and it decimated 20 percent of all life-forms. The third, around 250 million years ago, can be called the Big One. Recent discoveries in southern China and Australia strongly suggest that a comet or a meteorite smashed into the earth, punching through the mantle and liberating unfathomable volumes of sulfur, leading to a withering acid rain and "the most severe biotic crisis in the history of life on Earth."[15] This was a monumentally close shave for life on Earth. Only 5 percent of all marine life survived, and most of the land animals, and the plants they ate, were also wiped out.[16] But not everything died. Whatever survived was subjected to the fourth major extinction, around 200 million years ago, possibly a result of atmospheric changes due to massive volcanic activity. It wiped out close to half the species on Earth. But, again, many survived, and not long afterward the first dinosaurs appeared. But they were wiped out in the fifth major extinction, around sixty-five million years ago, when yet another comet or hunk of space rock slammed into Earth, devastating the dinosaurs, but opening an opportunity for the early mammals to thrive.[17]

## So Where's the Balance?

What's missing from both invasions and extinction events is a discernible plan of Nature (with a capital N), a plan to introduce a new species to an environment, or a plan to cull certain types, by extinction, in order to "balance" a system.

Instead, we see catastrophes that come from space, for example, meteors or comets that smash into Earth and nearly wipe out all life. Or we see climate fluctuation due to changes in how Earth rotates, fluctuation that probably caused the ice ages, which have wiped out thousands of species while allowing other thousands to flourish. Or we see continents on the move, driven by the gigantic forces of geology, creating land bridges that are eventually crossed by invasive species.

Arguing that these kinds of evolution-driving events are part of a great plan of Nature forces us to define what Nature is. If Nature is restricted to life on Earth, and some of the most profound changes to life on Earth result from space-debris impact, then we can reasonably ask, how does Nature import space debris for its *balancing* of life on Earth? How could earth-based Nature do that?

And if Nature is said to extend *beyond* Earth, does it include the entire universe? Where does it end? Technically speaking, it should include the entire universe. But then we come to an absurdity, because to attribute the entirety of the universe, and everything that happens in it, to a discrete plan with checks and balances would be such a nebulous "explanation" that it wouldn't really explain anything, it would only be "explaining away."

Finally, of course, if Nature—with a capital N—was intent on balancing things, it would have to have a consciousness with which to act out its wishes, and there is simply no scientific evidence for either intent or consciousness.[18]

Extinction events and invasive species, then, reveal the truth about ancient life systems. As much as we'd like to believe in benevolent forces that maintain perfect balance—promoting some species, while culling others—there just isn't evidence for them. There's much more evidence for chance events shaping the history of life on Earth.

But wait! Just because random events have effects on the trajectories of evolution, that doesn't mean that all is chaos, that order and balance can never occur. It's obvious that nature is full of balanced ecosystems. For example, we know that in a given ecosystem, if there are "too many" predators (and by definition, "too few" prey), some predators will starve and die, and the system—measured in the ratio of predators to prey—will stabilize, or balance out. But it's critical to remember that this isn't balance by intent: it's *equilibrium*, a concept we'll examine below. Order occurs, but this doesn't mean that nature itself has a purpose or goal. If it did, we'd have seen it in the fossil record of the 3.5 billion years of life on Earth. We see order and equilibrium there, because ecosystems do tend to stabilize over time, but we also see enormous, intent-free disturbances to those ecosystems.

Considering all the chaos we've seen, is there any use left in the phrase "the balance of nature"? Well, as we just saw, there *are* balances in nature; they just aren't a product of intent. The study of ecology is the study of the communities that achieve equilibrium, and of communities that are occasionally blasted out of it. Perhaps that field can clarify what balance in nature really means.

## Balance Concepts in Modern Ecology

We might think that "objective science" would avoid personifying the natural world, but scientists are humans, and they've made plenty of mistakes.[19] Since the fifth century BCE, when Greek historian Herodotus marveled at how nature maintained predator and prey populations by providing them with different birthrates, one of the principal mistakes of science has been the personification of nature, endowing it with intent, and even endowing its ecosystems and species—from entire jungles down to individual spiders—with specific functions to play in maintaining the balance.[20]

Although scientists don't personify nature as much as they used to, the concept of balance remains a foundation of both modern scientific ecology as well as popular ideas about the nature of the living world.[21] That's well and good because equilibrium does occur in evolving natural systems, but we have to remember it's not because of intent. Actually, several types of equilibrium occur. In his examination of the concept of *equilibrium* in living systems, ecologist Stuart Pimm found that this single word masks several meanings.[22] Although equilibrium (also known as *stability*) generally refers to a system's capacity to reorganize once disturbed, the word is alternately used to mean *resilience* (the *speed* at which a part of a system returns to stability), *persistence* (the *duration* of stability in a system), or *resistance* (the *magnitude of effects* that disturbances have on stability). Pimm goes on to point out that even professional ecologists often fail to recognize the real complexity of the concept of equilibrium at the heart of the phrase "the balance of nature." In fact, his book, published in 1991, is tellingly titled as a question—*The Balance of Nature?*—while an influential book published in 1960 more simply assumed, in its title, *The Balance of Nature*.[23]

## Does Evolution Maintain Nature's Perfect Balance?

We've learned two significant lessons in this chapter. First, although balance does happen in nature, it's not the *goal* of evolution—something that evolution is striving to achieve—because evolution doesn't have the intent to strive in the first place. Remember, evolution is a single word we use to describe the cumulative effects of the three somewhat independent processes of *replication*, *variation*, and *selection*, so evolution can't even think to make a plan, let alone carry out a plan.

Second, if evolution isn't working toward a goal of balance (although balance does happen in nature), then the members of coevolving natural systems don't have roles and, as such, they're not there *to do* anything. Even symbiotic relationships, like that between sea anemones and hermit crabs (the anemones get a free ride, and the hermit crab is protected from octopus attacks), aren't planned or even maintained by a natural intent. They simply occur. In time, every balanced system will break apart. And in time, the survivors will again settle into an ecological order.

There's an obvious difference between, on the one hand, an intentionally designed, functional system, maintained by actors fulfilling roles according to an intended plan, and on the other hand, a complex system in which order arises simply as a result of ecological laws. But it can be very hard *not* to imagine some intent in nature. That's probably because for so long humanity has purposively made things—from stone tools to space shuttles—*with an end in mind.* Maybe it's not surprising that we've wrongly assumed that nature has worked that way too.

But nature doesn't draw up a plan, or work toward an end. Evolution (replication, variation, and selection) simply happens, and life-forms settle into balanced ecosystems. When comets occasionally impact our planet, or when continents slide into each other, some species are wiped out and others survive. When order and complexity arise out of chaos, as they almost always do,[24] we often see a pattern and assume there must be a mind, a pattern-maker, behind it. But our examination of the fossil record and of the causes behind the rise and extinction of millions of life-forms over the last few billion years has shown that the major patterns of nature aren't the result of an intended plan. Rather, they represent the result of chance events that disturb equilibrium and of the reorganization of equilibrium by the survivors.

## NOTES

1. Catastrophes occur, of course, but on the human timescale, they're rare by definition. The term *Mother Nature* is first traced to 1601, and "Mother Earth" to 1586. See *Oxford English Dictionary*, 2nd ed., vol. 9 (Oxford: Oxford University Press, 1989), p. 1125.

2. Some kind of information was stored as notches on European bone and antler "batons" over fifteen thousand years ago. See F. D'Errico, "Palaeolithic Origins of Artificial Memory Systems," in *Cognition and Material Culture: The Archaeology of Symbolic Storage*, ed. C. Renfrew and C. Scarre (Cambridge, UK: MacDonald Institute for Archaeological Research, 1998), pp. 19–50. However, the first records we can actually read are the first written languages of about five thousand years ago. See S. D. Houston, *The First Writing: Script Invention as History and Process* (Cambridge: Cambridge University Press, 2004).

3. The term *coevolution* refers to organisms whose evolution is intimately tied to that of others. Symbiotic species, as well as parasites and hosts, for example, coevolve. The term can also be applied to larger systems of "coevolving community members."

4. D. Raup, *Extinction: Bad Genes or Bad Luck?* (New York: Norton, 1991), p. 55.

5. Ibid., p. 108.

6. Raup and Sepkoski once suggested that there was a twenty-six-million-year extinction cycle. See D. Raup and J. Sepkoski, "Periodicity of Extinctions in the Geologic Past," *Proceedings of the National Academy of Sciences USA* 81 (1984): 801–805. But by 1991 Raup noted that numerous studies of the same data set did not support this cycle and that no mechanism for it has been generally accepted. See D. Raup, *Extinction*, pp. 164–65.

7. The study of animal behavior is *ethology*. The behavior of species known only from the fossil record can sometimes be deduced with great accuracy, because anatomy, for example, can tell us about diet, and thus feeding patterns, seasonal patterns, and so on. *Form*, *function*, and *behavior* are directly related, and often readable. For fascinating examples, see C. Oxnard, *Fossils, Teeth and Sex: New Perspectives on Human Evolution* (Seattle: Washington University Press, 1987).

8. Most plant seeds regularly travel only a few meters (about ten feet), though some are regularly carried by wind as much as sixteen miles (twenty kilometers), and clearly some islands have been vegetated by seeds carried hundreds of miles either by wind or, often inadvertently (as in birds' stomachs), by animals. Coconuts can also survive for months at sea, travel enormous distances, and germinate far from home. See M. L. Cain, B. G. Milligan, and A. E. Strand, "Long-Distance Seed Dispersal in Plant Populations," *American Journal of Botany* 87, no. 9 (2000): 1217–27.

9. True *invasion* normally means that the arriving species produces fertile offspring at some distance from the point of entry into a new habitat. See D. M. Richardson et al., "Naturalization and Invasion of Alien Plants: Concepts and Definitions," *Diversity and Distributions* 6 (2000): 93–107. Invasive species are introduced in D. M. Lodge, "Biological Invasions: Lessons for Ecology," *Trends in Ecology and Evolution* 8, no. 4 (1993): 133–37; and M. Williamson, *Biological Invasions*, Population and Community Series 15 (London: Chapman and Hall, 1996).

10. For the evolutionary history and ecology of the marsupials, see J. A. M. Graves and M. Westerman, "Marsupial Genetics and Genomics," *Trends in Genetics* 18, no. 10 (2002): 517–21. For an overview, see A. K. Lee and A. Cockburn, *Evolutionary Ecology of Marsupials* (Cambridge: Cambridge University Press, 1985). The earliest known fossil marsupials are dated to just over 130 million years ago. See R. L. Cifelli and J. G. Eaton, "Marsupial from the Earliest Late Cretaceous of Western US," *Nature* 325 (1987): 520–22. The earliest eutherian (placental) mammal is dated to about one hundred million years ago. See Q. Ji et al., "The Earliest Known Eutherian Mammal," *Nature* 416 (2002): 816–22.

11. The "Great American Interchange" is nicely summarized in L. G. Marshall, "Land Mammals and the Great American Interchange," *American Scientist* 76 (1988): 380–88.

12. See Nature Conservancy, "Invasives Alert! *Anoplophora glabripennis* (Motchulsky): (Asian Longhorned Beetle)," http://tncweeds.ucdavis.edu/alert/alrtanop.html (accessed November 11, 2005).

13. Our dependence on honeybees to pollinate our food and commercial crops is so significant that it has been discussed in a New York Academy of Sciences study titled "Food and Agricultural Security: Guarding Against Natural Threats and Terrorist Attacks Affecting Health, National Food Supplies, and Agricultural Economics." See B. H. Thompson, "Where Have All My Pumpkins Gone? The Vulnerability of Insect Pollinators," in *Food and Agricultural Security: Guarding Against Natural Threats and Terrorist Attacks Affecting Health, National Food Supplies, and Agricultural Economics*, ed. T. W. Frazier and D. C. Richardson (New York: New York Academy of Sciences, 1999), pp. 189–98.

14. The mass extinctions can be reviewed at the American Museum of Natural History Web site. See "Humans and Other Catastrophes," http://

www.amnh.org/science/biodiversity/extinction/ (accessed October 8, 2005).

15. Quotation from K. Kaiho et al., "End-Permian Catastrophe by a Bolide Impact: Evidence of a Gigantic Release of Sulfur from the Mantle," *Geology* 29, no. 9 (2001): 815. For the Australian evidence for space-debris impact, see L. Becker et al., "Bedout: A Possible End-Permian Impact Crater Offshore of Northwestern Australia," *Science* 304, no. 5676 (2004): 1469–76. For a general account of the Permian extinction, see M. J. Benton, *When Life Nearly Died: The Greatest Mass Extinction of All Time* (London: Thames & Hudson, 2003).

16. For a textbook overview of extinction events, see R. Cowen, *History of Life*, 4th ed. (Malden, MA: Blackwell, 2005). Note that a lot of extraterrestrial threats remain, and NASA maintains an "Asteroid and Comet Impact Hazards" Web site. See http://impact.arc.nasa.gov/ (accessed January 22, 2005), where it reports on NASA's monitoring of hazardous space debris, including "civilization-killers," objects large enough to wipe out civilization.

17. Although there's debate on the cause of the fifth extinction, many believe it was the result of a comet impact. For example, species disappeared quickly rather than gradually. See P. M. Sheehan et al., "Dinosaur Abundance Was Not Declining in a '3 m gap' at the Top of the Hell Creek Formation, Montana and North Dakota," *Geology* 28, no. 6 (2000): 523–26. An impact site and associated evidence have been found near Mexico. See A. R. Hildebrand et al., "Chicxulub Crater: A Possible Cretaceous/Tertiary Boundary Impact Crater on the Yucatan Peninsula, Mexico," *Geology* 19, no. 9 (1991): 867–71. Still, there's room for debate, as reviewed in C. Officer and G. Page, *The Great Dinosaur Extinction Controversy* (New York: Addison-Wesley, 1996). Note that we're currently in the sixth extinction, one that's largely due to human activities. It's been estimated that, presently, three species go extinct per hour, a rate unseen for more than two hundred million years. For a detailed discussion of the sixth extinction, see N. Eldredge, "Cretaceous Meteor Showers, the Human Ecological 'Niche,' and the Sixth Extinction," in *Extinctions in Near Time: Causes Contexts and Consequences*, ed. R. D. E. MacPhee (New York: Kluwer Academic/Plenum, 1999), pp. 1–14; and R. D. E. MacPhee and C. Flemming, "Requiem Aeternum: The Last Five Hundred Years of Mammalian Species Extinctions," in *Extinctions in Near Time: Causes Contexts and Consequences*, ed. R. D. E. MacPhee (New York: Kluwer Academic/Plenum, 1999), pp. 333–72. For a more general account,

see R. Leakey and R. Lewin, *The Sixth Extinction: Biodiversity and Its Survival* (London: Weidenfeld & Nicolson, 1996). Continuing to destroy our plant's ecosystems could be suicidal since we depend on them for our very survival.

18. Suggesting that we simply can't *hope to understand* a grand Natural Plan is about the same as saying *it's all God's plan*, neither of which are useful statements in science, because once you bring in supernatural powers, all bets are off. The *Indiscernible Plan* and *God's Plan* "explanations" may be true (although there is no scientific evidence for either), but even so, neither would really *explain* the plan and would only identify that one exists. One could still ask *What is God's Plan?* or *What is Nature's Plan?* Saying these plans are eternal mysteries is the end of inquiry and unacceptable to the scientific system of knowledge.

19. The late astronomer and popularizer of science, Carl Sagan, often pointed out that scientists question traditional wisdom, and they even admit when they're wrong, changing their minds as new information disproves their previous beliefs. This way of generating knowledge is fundamentally different from simply acquiring knowledge by tradition, or simply accepting what people tell you. The motto of London's Royal Society, dedicated to the generation of new knowledge, rejects arguments from authority by its motto, *Nullius in Verba*, roughly meaning "by no man's word"; the Royal Society, and science in general, want evidence, not just words or declarations. For a general overview of the scientific method, with plenty of examples, see C. Sagan, *A Demon-Haunted World: Science as a Candle in the Dark* (New York: Ballantine Books, 1996).

20. The closest anyone comes to suggesting that Earth has a systemic consciousness are proponents of the Gaia hypothesis. This is the idea that Earth is a giant, self-regulating, living organism that actually seeks to maintain "optimal" conditions for life. Suzuki points out that the Gaia hypothesis wasn't formulated as possessing consciousness. See D. Suzuki, *The Sacred Balance: Rediscovering Our Place in Nature* (Toronto: Greystone Books, 2002), p. 145. But in popular media there is an undercurrent attached to the idea that suggests consciousness. Gaia is a fascinating concept, but it's also nebulous and has little scientific evidence to support it. When a chief proponent of a theory finds it difficult to describe systematically (as did Lynn Margulis, a cofounder of the Gaia hypothesis, as recently as 2000), and a world authority on ecological equilibrium (Stuart Pimm) chooses not to discuss

Gaia because he admits that he does not understand it, the theory is in trouble. See, respectively, J. Turney, *Lovelock and Gaia: Signs of Life* (New York: Columbia University Press, 2003), pp. 245–46; and S. Pimm, *The Balance of Nature? Ecological Issues in the Conservation of Species and Communities* (Chicago: University of Chicago Press, 1991), p. 5. Here, we're following Pimm's lead. See J. Turney, *Lovelock and Gaia: Signs of Life* for a recent review of Lovelock's Gaia concept. The Gaia hypothesis is not impossible, but it has yet to be substantiated, or even clearly defined.

21. For a scholarly review of the *balance of nature* metaphor in science, see K. Cuddington, "The 'Balance of Nature' Metaphor and Equilibrium in Population Ecology," *Biology and Philosophy* 16 (2001): 463–79.

22. See S. Pimm, *The Balance of Nature?*

23. See L. J. Milne, *The Balance of* Nature (New York: Knopf, 1960).

24. For a fascinating discussion of complexity and "self-organization" in nature, see I. Prigogene and J. Stengers, *Order Out of Chaos: Man's New Dialogue with Nature* (New York: Bantam Books, 1984).

## Myth Eight

# Creationism Disproves Evolution

In 1650 the Anglican archbishop James Ussher (1581–1656) set out to determine the age of Earth, but his approach was very different from that of modern geologists. He used the chronology and genealogies in the Bible to arrive at his answer.[1] Considering the margin for error involved in making such a calculation, it's surprising that Ussher narrowed it down to the day. He calculated that God created the world on the evening of October 22, 4004 BC.[2]

Not long after Archbishop Ussher made his calculation, scientific knowledge began to progress rapidly in western Europe. Scientific organizations were established, including the esteemed Royal Society of London in 1660, and the Académie des Sciences in France in 1666. It was in this Age of Enlightenment that Isaac Newton (1643–1727) discovered the laws of motion, and developed his theory of optics and of differential and integral calculus; the prolific Comte de Buffon (1707–1788) attempted to include everything known about the natural world in his staggering thirty-six-volume encyclopedia; and the inventor James Watt (1736–1819) developed the first practical steam engine.[3] Moreover, the refinement of lenses for microscopes and telescopes opened scientists' eyes to the infinitely small organisms around them and the infinite reaches of space.

Industry and trade also were on the rise at this time, and many canals for transporting goods and raw materials were being built across western Europe. Deciding where to build a canal requires knowledge of the earth's geological composition, primarily because certain kinds of rock and soil are better at holding water than others. A great many fossil discoveries were made during the building of these canals, especially in England, where the geologist William Smith (1769–1839) noticed that the relative placement of various fossils in the geological strata was essentially the same throughout distant parts of England. This consistency of layering needed some kind of explanation.

Up until this time, the accepted explanation for the existence of mountains, valleys, and other geological features was that they had been formed by violent, cataclysmic events in the past, events that drastically and suddenly altered Earth's terrain. This was the theory of *catastrophism*, and it suggested that natural catastrophes such as floods, earthquakes, and volcanoes were primarily responsible for Earth's features. This theory was popular with many biblically minded scholars because they interpreted it to mean that the catastrophic flood of Noah's time was responsible for Earth's present terrain. Catastrophism seemed to fit nicely with the Bible, and it lent support to Archbishop Ussher's "young Earth" calculation.[4]

But not everyone was convinced by the theory of catastrophism. The Scottish geologist James Hutton (1726–1797) offered an alternative explanation. Catastrophes, he suggested, weren't the main events that had changed the landscape. Rather, the forces that created geological features are the ones we see all the time, such as erosion, sedimentation, and glaciation. Given enough time, streams could shape valleys, sediments could accumulate to form new landmasses, and glaciers could cut through solid rock. Hutton reasoned that if these forces had been acting in roughly the same way since Earth's beginning, then Earth had to be very ancient, much older than Ussher calculated. This notion of slow, regular geological change was called *uniformitarianism*, and it was contrasted with the cataclysmic explanation of catastrophism.

The geologist Charles Lyell (1797–1875) published *Principles of Geology* in 1830, popularizing Hutton's theory of an ancient Earth. His arguments for uniformitarianism were so convincing that most scientists and religious thinkers gave up on the catastrophist explanation, whether or not they based it on the biblical account of Noah's flood.[5] Charles Darwin found many of Lyell's ideas helpful in providing evidence for his own views on evolution. This is because evolutionary changes in species happen so slowly that Darwin's theory needed an ancient Earth. The marriage of evolution and geology was consummated nearly thirty years later when Darwin published *On the Origin of Species*.

The theory of evolution predicts that if life evolved slowly over time, we should observe certain things in geological formations. The deeper we dig, for example, the more different from modern life the fossils will appear. And if life started out simple, then we should see simpler organisms farther down in the rock strata. If certain orders of animals—such as mammals—arose relatively late, then we should find no evidence of mammal fossils below a certain point in the geological strata. These and many other predictions have turned out to be true, and discoveries in geology and paleontology have provided overwhelming evidence supporting evolution. And geologists have shown, with evidence collected from around the globe, that Earth is roughly 4.5 billion years old.

Of course, members of certain religious sects have never been convinced by the overwhelming evidence for an ancient Earth. In the mid-twentieth century, particularly in America, there was a revival of the idea that Earth was not so ancient after all.

## Young Earth Creationism

For the Catholic and mainline Protestant Churches, evolution isn't seen as incompatible with their theology, and they've fought hard against the teaching of creationism in public schools.[6] Creationists themselves come in a number of varieties, and not all of them think Earth is young, but they do agree in their rejection of evolution.[7] The most vocal opponents of evolution are those Christian fundamentalists known as *Young Earth Creationists*. Their species of creationism was dramatically resurrected on the American stage with the publication in 1961 of the book *Genesis Flood*, by engineer Henry M. Morris and theologian John C. Whitcomb. Two years later, Morris founded the influential Institute for Creation Research in order to combat the teaching of evolution in public schools.[8]

Those at the institute rejected evolution and the view that Earth is old. They also dismissed the 150 years of data accumulated by geologists and paleontologists worldwide because those findings had been used to confirm evolution and the great antiquity of Earth. Creationists are fond of looking to the Bible to support their existing beliefs, and that's nothing new. What made those at the Institute for Creation Research unique was their attempt to muster rational-sounding arguments against evolutionary theory and, in particular, against the accepted scientific views of geology and paleontology that supported evolution.

The standard position of Young Earth Creationists, as put forward by Morris, is that Earth is somewhere between six and ten thousand years old. These creationists also believe that Earth (indeed, the universe itself) and all living "kinds" were created in six literal days, ruling out the possibility of the slow evolution of species over time.[9]

Ultimately, their evidence for these views is found in the Bible, which they consider "inerrant." In their view, the Bible is free from error and should be understood literally, not just in moral and spiritual matters, but even in descriptions or accounts of the physical world and its history.[10] This leads Young Earth Creationists to believe, among other things, that the actual physical body of Eve was made from one of Adam's ribs, and that Noah and his three sons actually built an ark capable of carrying two of each of Earth's creatures to survive the great flood. When asked to explain the existence of fossils and their unique layering, many Young Earth Creationists say they're simply the remains of dead creatures drowned in Noah's flood.[11] But this doesn't explain why we find the fossils of more complex animals higher in the rock strata. Some creationists even go so far as to claim that God placed the fossils there to test our faith.

For Young Earth Creationists, whenever the words of the Bible and the findings of science differ, the truth must always be found in the Bible. They're fond of scientific discoveries that can be interpreted to support the word of God as revealed in the Bible—not that their faith requires this support—but in the end, the Bible is the ultimate authority, and it trumps all scientific evidence by default. According to Morris and Whitcomb, there's no real choice when the Bible and the findings of geologists are at odds:

> The decision must then be faced: either the Biblical record of the Flood is false and must be rejected or else the system of historical geology which has seemed to discredit it is wrong and must be changed. The latter alternative would seem to be the only one which a Biblically and scientifically instructed Christian could honestly take. . . .[12]

Compared to the evidence of geology, the claims of Young Earth Creationists are so ridiculous that no legitimate scientists believe them, and these creationists haven't been successful at introducing their views into the public schools in the United States. The establishment clause of the First Amendment to the Constitution stands in their way, stating in part that "Congress shall make no law respecting an establishment of religion." To sidestep this impediment, Young Earth Creationists have tried to establish what they call "creation science." The goal of creation science is to dispute scientific claims or theories that are at odds with a literal interpretation of the Bible by trying to argue against them on scientific grounds. This approach, they think, puts creation science on par with mainstream science, in that it simply offers alternative scientific explanations (or disproves existing scientific explanations). If this approach works, they believe they have a good chance of making Young Earth Creationism—and a rejection of evolution—a part of the science curriculum in public schools.

But creation science hasn't worked because it can't stand up to even very simple examination. So, instead of coming up with their own scientific explanations, these creationists focus on the supposed flaws in the scientific evidence for an old Earth, and by extension, evolution. For example, Young Earth Creationists say that if evolution can be shown to be mistaken or flawed, then creationism must, by default, be the truth. But this is a flawed argument.[13] Even if evolutionary theory was falsified, it wouldn't follow that creationism would then be true. There might be other explanations for the fossil record, and for the variety of species on the planet, which have nothing to do with evolution or creationism.

If Young Earth Creationism is ever going to compete with mainstream science, it will have to do much better than this. But, so far, creationists haven't offered any real evidence for their views, and that's because there isn't any evidence for those views.

## Gaps for God

What are the flaws in the theory of evolution, according to creationists? To consider all of the objections would require an entire book on creationism. But let's look at a few of their more popular views.

Many creationists claim that if evolution were true there should be transitional fossils between species, but since there are no transitional fossils, evolution must be false. This is simply wrong. Notwithstanding the odds against fossilization, there are plenty of transitional fossils, despite creationists' claims. These transitional fossils include the archaeopteryx, a crow-sized animal that had features of both reptiles and birds. There are australopithecine fossils, which are transitional between apelike animals and humans. And most famously there are the many transitional fossils leading from *Eohippus*, the earliest known member of the horse group, to modern horses. In fact, there are so many known transitional fossils between reptiles and amphibians,[14] as well as between an order of warm-blooded reptiles (called *therapsids*) and mammals,[15] that it's hard to know exactly which groups to classify them in.

There are gaps in the fossil record, of course. How could it be complete? Soft-bodied animals such as worms will rarely fossilize. Hard body parts such as shells, bones, and teeth are more likely to fossilize, but most don't because they've either been broken into bits, chemically dissolved, weathered, or they just didn't end up where they needed to be, namely, where sedimentary rocks formed. Even if specimens do become fossilized, many are easily destroyed before they're ever found because of erosion, shifting landmasses, earthquakes, and volcanoes. The earth can change a lot in millions or hundreds of millions of years. Yet creationists are fond of pointing out the existence of gaps in the fossil record, as if this disproves evolution and confirms creationism. This is like arguing that if you can't trace your ancestry back more than, say, two hundred years, that you have no ancestors older than those, and that those ancestors must have popped up out of nowhere in some special act of creation.

## The Dating Game

Young Earth Creationists also argue that radiometric dating is flawed. Radiometric dating determines the age of rocks or organic remains by measuring the decay of naturally occurring radioactive elements. But Young Earth Creationists argue that the science of radiometric dating is unreliable. The rates of radioactive decay, some claim, are too poorly known to be used accurately. But this is just flat-out false. Throughout the last hundred years the decay rates of many radioactive isotopes have been thoroughly measured. In some cases the materials are precisely weighed, then left to sit for years, and then weighed again in order to determine their decay rates. Radioactive decay can also be measured accurately with Geiger counters and gamma-ray detectors to determine the energy bursts given off by the decaying radioactive material.[16]

Many creationists enjoy citing the case of a living freshwater mussel that was once measured by carbon-14 dating to be over two thousand years old. They argue that if a living mussel was dated to be over two thousand years old, then radiometric dating is surely flawed and must be abandoned. But this incorrect date is easily explained by showing that there is very little carbon 14 in the mussel's environment due to the age of the limestone in the water, and the small amount of atmospheric carbon that can penetrate the water.[17]

The hopelessness of this creationist argument is pointed out by the renowned geologist G. Brent Dalrymple, who offers an apt reply: if you'd bought a watch that didn't keep good time, you wouldn't assume that all watches were flawed. You'd assume that there was an explanation for why your watch wasn't working properly.[18] This is the same for radiometric dating. Different techniques are only useful under certain geological conditions. And various factors—such as the mussel's unique environment—have to be considered. Sometimes mistakes are made, and then they are investigated and corrected. This is how good science works. Regardless of creationists' claims, the accuracy of radiometric dating is so well established that no serious geologist or paleontologist doubts it.

## From Bad to Weird

There are a host of bad arguments that Young Earth Creationists drag out from time to time. These include, among others, the claim that evolution violates the second law of thermodynamics; that rare examples of more complex fossils beneath simpler fossils completely refute evolution; that there should be larger quantities of salt left over from evaporation if Earth is as old as geologists say; that the rate of decay of Earth's magnetism proves Earth is only ten thousand years old; that there should be a hundred-foot layer of meteorite dust covering the surface of Earth and the moon if they're both 4.5 billion years old; and that dinosaur and human footprints supposedly found together in Texas prove that humans and dinosaurs lived at the same time—what zoologist Tim Berra calls the Fred Flintstone version of prehistory.[19] But every one of these arguments fails to be supported by scientific evidence.

Creationists have made other bizarre claims in order to persuade the public (as if the above examples aren't bizarre enough), but they have wisely abandoned many of them. These include the claim that women have one more rib than men; that the universe appears old because light used to travel faster a few thousand years ago; and that when NASA computers were calculating the position of the planets, they discovered the "missing day" as mentioned in the Bible (Joshua 10), when God made the sun and the moon stand still. All of these claims by creationists are just simply wrong.

Creation science has ultimately failed the test of science because it doesn't offer any compelling scientific evidence, and in every case so far its rejection of existing scientific explanations that support evolution has been erroneous. Creation science has also failed the test of the courts because it's obvious that its positions are not scientific, but religious.[20] Since their attempts to get creation science taught in the public schools have failed, many creationists of different stripes have joined forces and formulated a more clever strategy. This strategy, which they hope will past the tests of the courts, is known as *intelligent design*.

## Notes

1. Archbishop Ussher's book published in 1650 was titled *Annales Veteris Testamenti, a Prima Mundi Origine Deducti* (*Annals of the Old Testament, Deduced from the First Origins of the World*).

2. A few years after Archbishop Ussher's calculation of the date of creation, Cambridge University's John Lightfoot (1602–1675) calculated the date as October 18, 4004 BC, and the creation of Adam at 9:00 AM, on October 23.

3. Gottfried Wilhelm Leibniz (1646–1716) also developed calculus, independently of Newton. Georges-Louis Leclerc, Comte de Buffon (1707–1788), also suggested that Earth was older than six thousand years, and that humans and apes might share a common ancestry.

4. Not all catastrophist geologists were dogmatic adherents to the Noachian flood explanation. See S. J. Gould, *Time's Arrow, Time's Cycle* (Cambridge, MA: Harvard University Press, 1987). The French naturalist Baron Cuvier (1769–1832) was a catastrophist who argued that a great many catastrophic events in the past were responsible for existing land formations. He, as much as Lyell, persuaded many to reject the idea of a single worldwide deluge as an explanation. See R. J. Shadewald, "The Evolution of Bible-Science," in *Scientists Confront Creationism*, ed. L. R. Godfrey (New York: Norton, 1983), p. 286.

5. Modern geologists recognize that there have been numerous cataclysmic events in Earth's history, but that these cataclysms are viewed against a background of geological uniformitarianism.

6. In the court case *McLean v. Arkansas Board of Education* (1982), the plaintiffs (those *against* teaching creationism in the Arkansas public schools) included Arkansas bishops of the Roman Catholic, United Methodist, Episcopal, and African Methodist Episcopal Churches, as well as the head official of the Presbyterian Church in Arkansas, and other Presbyterian, United Methodist, and Southern Baptist clergy. Evolution is taught without any creationist twists in such respected religious-affiliated universities as Notre Dame (Catholic), Brigham Young (Mormon), and Baylor (Baptist).

7. Old Earth Creationists, as the name suggests, believe Earth is old, and many interpret the six days of creation in Genesis to represent long ages rather than literal days (Day Age Creationism), while others believe there is

a long time gap (Gap Creationism) between Genesis 1:1 (when God created the Heaven and the Earth) and Genesis 1:2 onward, when the rest of creation arises. Old Earth Creationists, however, typically reject evolution.

8. After the infamous Scopes Monkey Trial in Tennessee in 1925, many high schools, particularly in the southern states, omitted any mention of evolution in their biology textbooks. But in 1957 the Soviet Union launched the first artificial satellite, *Sputnik*, into orbit. In response, the US government made efforts to revamp science education, which included developing up-to-date science textbooks, and it was at this point that evolution found its place in high school biology textbooks. It was this introduction of evolution that precipitated the backlash by creationists.

9. The word *kinds* is a biblical, not scientific, term. And many creationists believe there can be variation within kinds, but that no new kinds can evolve. It's not clear whether creationists take the word to mean species, genus, family, or order.

10. Young Earth Creationists recognize that there are passages in the Bible that contain parables, metaphors, and poetic images, but they maintain that these passages are clearly understood not to be taken literally.

11. H. M. Morris and J. C. Whitcomb, *The Genesis Flood* (Philadelphia: Presbyterian and Reformed Publishing, 1961), p. 123.

12. Morris and Whitcomb, *The Genesis Flood*, p. 118.

13. For a clear account of how creationists often rely on the either/or approach, see R. T. Pennock, *Tower of Babel: The Evidence against the New Creationism* (Cambridge, MA: Bradford Books/MIT Press, 1999), chap. 4.

14. T. M. Berra, *Evolution and the Myth of Creationism: A Basic Guide to the Facts in the Evolution Debate* (Stanford, CA: Stanford University Press, 1990), p. 127.

15. E. Mayr, *What Evolution Is* (New York: Basic Books, 2001), p. 14.

16. E. C. Scott, *Evolution vs. Creationism* (Westport, CT: Greenwood, 2004), p. 155.

17. Berra, *Evolution and the Myth of Creationism*, p. 131.

18. G. B. Dalrymple, "How Old Is the Earth? A Reply to 'Scientific' Creationism," in *Evolutionists Confront Creationism: Proceedings of the 63rd Annual Meeting of the Pacific Division, American Association for the Advancement of Science*, vol. 1, pt. 3, ed. F. Awbrey and W. Thwaites (AAAS, 1984), pp. 76–77.

19. Berra, *Evolution and the Myth of Creationism*, p. 129. Only 48 percent of Americans polled in 2001 knew that humans and dinosaurs never lived at the same time. See "Science and Technology: Public Attitudes and Understanding," in *Science and Engineering Indicators* (Arlington, VA: National Science Foundation, Division of Science Resources and Statistics, 2004), chap. 7, pp. 15–16, http://www.nsf.gov/statistics/seind04/pdf/c07.pdf (accessed May 5, 2005).

20. Eight significant court cases where creationists have been unsuccessful include: *Epperson v. Arkansas* (1968), *Segraves v. State of California* (1981), *McLean v. Arkansas Board of Education* (1982), *Edwards v. Aguillard* (1987), *Webster v. New Lenox School District* (1990), *Peloza v. Capistrano School District* (1994), *Freiler v. Tangipahoa Board of Education* (1997), and *LeVake v. Independent School District* (2001).

## Myth Nine

# Intelligent Design Is Science

On the first page of his book *Natural Theology*, the British theologian William Paley (1743–1805) imagines that he's walking through a field and stumbles over a stone. If someone were to ask him where the stone came from, his answer would be that for all he knows, it has been there forever. Then Paley imagines finding a watch on the ground, and entertains the question of where the watch came from. The answer, he tells us, couldn't be the same he gave for the origin of the stone. This is because the order and organization of the watch, unlike the stone, clearly shows that it was made by someone, that it was designed with a purpose in mind. This intricate object couldn't have simply come about by chance. Paley points out that much in nature is also orderly and organized, from the orbits of planets to the organs of animals. And he argues by analogy that if the order and organization of a watch is evidence for a designer with a purpose, then the order and organization found in nature is also evidence for a designer with a purpose.[1] Moreover, Paley insists that a thing as organized as the vertebrate eye—like the watch—couldn't have arisen by chance and must have been created by God, the ultimate designer.

Darwin was well aware of Paley's famous argument from design, and it's no coincidence that he used the eye as an example of an organ that could indeed evolve by natural selection, as opposed to being the product of a designer with a purpose. Darwin was not trying to disprove the existence of God. Rather, he was simply showing that the complex eyes of vertebrates could evolve by his theory of "descent with modification." Although he realized that this might sound absurd, he went on to say:

> Yet reason tells me, that if numerous gradations from a . . . complex eye to one very . . . simple, each grade being useful to its possessor, can be shown to exist; if further, the eye does vary ever so slightly, and the variations be inherited, which is certainly the case; and if any variation or modification in the organ be ever useful to an animal under changing conditions of life, then the difficulty of believing that a . . . complex eye could be formed by natural selection . . . can hardly be considered real.[2]

Creationists often ask, "What's the use of half an eye?" as if a lesser-quality eye would be of no use to the animal that possessed it.

But it's not a matter of all or nothing. If you're nearsighted, it's obvious that some vision (bad as it is) is better than none. A simple visual receptor can be useful, as Darwin points out. And it can certainly provide an advantage in avoiding predators, and in finding food and mates. We can even observe varying degrees of the optical quality of eyes in the Animal Kingdom, from the simplest lensless light detector, as found in the mussel-like nautilus, to the keen eyes of eagles and hawks.[3]

But nearly 150 years after Darwin's *On the Origin*, Intelligent Design Creationists are still peddling Paley's argument from design, although they've dressed it up with more contemporary examples and language. And as we'll see, this new version runs into the same problems as Paley's argument.

## REDUCE THIS

The modern intelligent design movement got its start in 1984 with the publication of *The Mystery of Life's Origin* in which the authors maintained that the beginning of life on Earth could never be explained by natural causes.[4] This was followed in 1989 by the book *Of Pandas and People*, where the phrase "intelligent design," or ID, first appears.[5] But it wasn't until 1993 with the publication of Phillip Johnson's *Darwin on Trial* that ID began to attract public attention. As a law professor at a top-tier institution (University of California at Berkeley), Johnson lent academic credibility to ID. But Johnson has mostly been a publicist, whereas biochemist Michael Behe and philosopher/mathematician William Dembski have tried to present ID as a legitimate scientific theory. Both Behe and Dembski insist that some organisms—or parts of organisms—are so complex that the slow, cumulative processes of evolution could not have produced them. Instead, they claim to show scientifically that certain complex biological features must have been designed by an intelligent being.

The main argument of Michael Behe's book, *Darwin's Black Box*, is that some biochemical structures in nature are so complex that they couldn't have evolved by natural selection. He calls these structures *irreducibly complex*. For Behe, an irreducibly complex structure is composed of three or more interconnected parts, each of which must be present for it to function.[6] Behe claims that these parts could not have evolved by the cumulative process of natural selection, where parts of a complex structure are built up, piece by piece, over long periods of time.[7] Rather, he maintains that these structures must, by definition, have been designed (by a designer) as a functioning whole, because if one of the parts is removed, the structure itself fails to work.[8]

Unlike Paley, Behe is not as concerned with eyes or with the watch analogy. He prefers the analogy of a mousetrap to explain irreducible complexity. A mousetrap has numerous parts that work together in order to function. Because the parts are interconnected, if you remove one part the mousetrap won't function. This, Behe claims, is proof that the mousetrap had to be designed by an intelligence. Behe extends his analogy to nature, just as Paley did, but his main example is the tail that certain bacteria use for locomotion, called a flagellum.

Behe's approach to identifying the hand of a designer in nature is not much different than Paley's. But the solution to Behe's challenge is much the same as Darwin's solution to Paley's challenge, namely, that complex structures can evolve by natural selection. The biologist H. Allen Orr explains that:

> An irreducibly complex system can be built gradually by adding parts that, while initially just advantageous, become—because of later changes—essential. The logic is very simple. Some part (A) initially does some job (and not very well, perhaps). Another part (B) later gets added because it helps A. This new part isn't essential, it merely improves things. But later on, A (or something else) may change in such a way that B now becomes indispensable. This process continues as further parts get folded into the system. And at the end of the day, many parts may all be required.[9]

As we'll see below, there are a number of processes which, by means of gradual evolutionary change, can account for what Behe calls irreducible complexity.

First, molecular biologists know that genes sometimes make extra copies of themselves, copies the organism can survive just fine without. But over the long course of evolution these duplicated genes can change and come to perform new but related functions, which are necessary for survival. We have many genes that began as duplicates but are now a necessity, such as *myoglobin*, which transports oxygen in the muscles, and *hemoglobin*, which transports oxygen through the blood.[10]

Second, duplicated pieces of genes can combine together to form new genes with different functions. We can see this when the same parts of genes are combined differently to form the proteins involved in blood clotting, as well as in the digestive enzyme *trypsin*.[11]

Third, certain animal organs that weren't initially essential for survival (but did provide an advantage) can become essential. A case in point is the evolution of early lungs from air bladders, which allowed water-dwelling animals to explore the land. Those lungs weren't necessary for survival, yet they were an advantage. But lungs are now essential for land animals (not simply an advantage).[12]

These are just some examples of how changes that are not essential initially can become necessities over time. So, what Behe calls "irreducibly" complex systems are not irreducible. They can evolve over time by natural means, without intelligent design.

Behe would certainly admit that there's much we don't know when it comes to evolutionary biology. But his mistake is in implying that if scientists don't have a detailed answer for the evolution of every organ and body part of every single organism, then evolution must not be the best explanation. Evolutionary science certainly has many unsolved questions, but that just means we haven't yet figured out every detail of the step-by-step evolution of such complex features as the bacterial flagellum. Most scientists would think this signals a need for further research. Not Behe. He implies that, due to our ignorance, we should declare an end to further research and posit an intelligent designer instead, arguing that we can't conceive of the gradual evolution of such features.[13] But it's too easy to simply declare that we can never understand certain aspects of the natural world because they're too complex. This defeatist approach closes the door on future discoveries. Where would we be in our understanding of the natural world today if the likes of Galileo, Newton, Einstein, and Darwin had given up explaining complex problems and instead simply lifted their hands in defeat, declaring an intelligent designer as the best and only explanation? We would be stuck back in the Dark Ages, that's where.

## The Wedge by Design

So why do Behe and other ID proponents get so much press coverage, despite their flawed arguments? It has a lot to do with the efforts of the Discovery Institute and its religious-based Center for Science and Culture (CSC) in Seattle. In the late 1990s the CSC developed its *wedge strategy*. This included a five-year plan to publish thirty books and a hundred scientific papers in respected peer-reviewed journals. But perhaps a five-year plan was a bit ambitious, since no article supporting ID has yet to appear in any respected, peer-reviewed scientific journal. The wedge strategy also involved swaying public opinion in favor of ID and taking legal action to get it taught in public schools.[14]

Unlike traditional creationists, Intelligent Design Creationists are more sophisticated, generally avoiding references to God or the Bible in their works. This is how they propose to leap the hurdle of the establishment clause of the First Amendment, the clause that has tripped up traditional creationists. When ID advocates are asked who the intelligent designer is, they often claim ignorance, saying only that there must be a designer who is intelligent. When asked about their personal (rather than scientific) beliefs, many, including Behe, say they think the designer is God.[15]

Since ID proponents mustn't mention God if they're ever to get

ID into public schools, they've been forced into absurdities, such as admitting that space aliens could be the intelligent designers. Incidentally, this is a view held by a sect of believers in intelligent design called Raëlians, who claim that humans were created by extraterrestrials who are masters of genetic engineering.

Ironically, the wedge approach has driven a wedge between traditional creationists and Intelligent Design Creationists. Traditional creationists can't stomach such talk of space aliens as the intelligent designers, for the obvious reason that they believe God is the designer. But many bite their tongues, knowing that ID may be their last and only hope of getting their creationist foot in the public school door. And they know that talk of God just won't cut it with the courts.

So ID attracts a motley cast of characters, from Young Earth Creationists to those like Behe, who accept that humans and apes have a common ancestor,[16] and many in between. Needless to say, the discussions must get heated at the ID family reunion when all the relations are there.

The result of wanting to appease traditional creationists, while also trying to pass themselves off as scientific, is that ID spokespeople often speak out of both sides of their mouths concerning where they stand scientifically and theologically.

For example, in *Of Pandas and People*, authors Percival Davis and Dean Kenyon avoid mentioning faith or God. They play the political game well, referring only to such ill-defined notions as "intelligent design," "intelligent causes," and "a master intellect."[17] But earlier, Davis coauthored a book called *A Case for Creation*, in which his theological views toward science are clear:

> We accept by faith the revealed fact that God created living things. We believe God simultaneously created those special substances (nucleic acids, proteins, etc.) that are so intricately interdependent in all of life's processes, and that He created them already functioning in living cells.[18]

Davis's "scientific" view is designed to conform to his already existing religious beliefs. But that's not how science works. If your scientific hypothesis first has to pass through the filter of your religious faith, you risk losing any semblance of impartiality and objectivity. And those qualities are essential to good science, because scientific positions are subject to revision and even rejection if there's compelling evidence against them. Among some religious believers, though, it's a virtue to hold onto an article of faith even in the face of overwhelming evidence against it. A historical example of an extreme faith commitment comes from Saint Ignatius of Loyola (1491–1556), founder of the Catholic Society of Jesus (Jesuits), when he wrote, "To be right in everything, we ought always to hold that the white which I see, is black, if the Hierarchical Church so decides it. . . ."[19] Whether the filter is the Christian Church, the Bible, the Torah, or the Koran, the risk of sacrificing objectivity is itself a convincing reason to avoid mixing faith and science.

This doesn't mean that scientists can't also be people of faith. There's a particular theological position called *theistic evolution* that says one can accept the scientific understanding of evolution while maintaining one's belief in God. In this view, God does not miraculously tinker with the natural world, making adjustments here and there. Rather, God created the universe and set it in motion with all of its natural laws in operation. He doesn't need to fine-tune the long unfolding of creation—including biological evolution—through divine intervention. In other words, he doesn't give the leopard its spots, or the platypus its bill. Theistic evolution allows scientists to proceed as normal, keeping miracles out of the practice of science. Curiously, ID philosopher/mathematician William Dembski has pointed out that God's creation need not miraculously violate the laws of nature.[20] This seems like theistic evolution, and here Dembski doesn't sound like a creationist who insists on God making divine course corrections. However, Dembski has stated, in no uncertain terms, that "design theorists are no friends of theistic evolution."[21]

As we can see, when ID promoters try to appeal to both mainstream Americans and fundamentalist creationists, they can't help but change their story depending on their audience. Ultimately, they can't disguise that ID is not science, but just another form of creationism. It's no wonder that one book examining ID is titled *Creationism's Trojan Horse*.[22]

## All Natural

There's another problem with ID when it comes to understanding how science actually works. Because of their religious beliefs, ID proponents are at odds with the view that there's nothing more beyond the natural or physical world. This view, called *philosophical naturalism*, rejects the existence of anything that can't be explained—at least in principle—by natural laws or by purely natural methods of investigation.[23] Philosophical naturalists deny that there can be such things as immortal souls, angels, and even God, since none of these can be investigated scientifically. Some ID proponents claim that if we rely strictly on the natural methods of science, we're forced to deny the existence of souls, angels, and God. But this is not the case. All that's required of scientists is that they recognize that science is limited to using natural methods and natural explanations for what it studies. This view is called *methodological naturalism*, and it has nothing at all to say about things that can't be investigated using natural methods, such as souls, angels, or God.[24]

The difference between these two forms of naturalism is important for believers because it allows them to accept that science should be practiced with strictly natural methods, but it doesn't require them to accept the philosophical position that there is nothing beyond the natural world. Things that transcend the natural world—such as souls, angels, and God—may exist. But discovering them is not something science is suited for. Ultimately, science can't tell us if God, souls, or angels exist or not—if we assume that these are not part of the physical world.[25] And we shouldn't expect to find God's "footprints" by peering through a microscope at a bacterial flagellum, just as we shouldn't expect to locate the kingdom of heaven by gazing at the night sky through a telescope.

But both Dembski and Behe consider it a mistake for science to limit itself to natural causes and explanations.[26] In fact, Behe's understanding of what counts as a scientific theory is so broad that it would include astrology as science.[27]

If science opens the door to nonnatural or divine explanations, the reliability and integrity of science itself is at stake. Part of the strength of science is in its universality, its ability to offer explanations that can be tested and examined by many scientists regardless of their theological or philosophical views, or whether they come from Tokyo, Tehran, or Toronto. Theology, however, is not testable in this way, which makes theological and scientific claims of a completely different nature. This is why a person's theological views have no bearing on the fact that a water molecule consists of two atoms of hydrogen and one atom of oxygen.

It's no secret that ID proponents try to distance themselves from creationism when addressing mainstream audiences. So it's rather surprising that Phillip Johnson is a chief spokesman for ID since his creationist agenda is so clear. He says that the proper basis for science is not naturalism, and he offers the following passage from the Gospel according to John as the preferred basis for science.[28]

> In the beginning was the Word, and the Word was with God, and the Word was God. He was in the beginning with God; all things were made through him, and without him was not anything made that was made. (John 1:1–3)[29]

How we're supposed to interpret this passage as forming the new basis for science is utterly baffling. And while the rest of us scratch our heads in confusion, Johnson eagerly awaits the fall of naturalism as he prepares for the Gospel of Christ to takes its place.[30] This vision of what Johnson calls "theistic science"—where divine explanations and interventions become part of the scientific method—ought to be disturbing to anyone who values science as an attempt to explain the natural world as impartially and objectively as possible.[31] An article in the Vatican newspaper, *L'Osservatore Romano*, recognized the dangers of confusing science with religion and came down clearly against ID, stating: "Intelligent design does not belong to science and there is no justification for the pretext that it be taught as a scientific theory alongside the Darwinian explanation."[32]

So, we can see that belief in God is not incompatible with evolution or science in general.[33] One can simply accept that science works best by limiting itself to natural explanations. But once we inject nonnatural or divine explanations into the practice of science, we're not really doing science anymore. We're either doing theology or groundless speculation in the guise of science.

## What Controversy?

A clever move by ID proponents is to claim that they just want to "teach the controversy" of evolution in public schools. On the surface this sounds very fair-minded and objective. Teaching the controversy might have saved Galileo a lot of grief when he argued that the sun, not Earth, is at the center of our planetary system. But we wouldn't "teach the controversy" between a sun-centered and earth-centered system in astronomy classes today, simply because there is no controversy. The sun-centered system has been accepted because the evidence shows it to be a fact. The same holds for evolution. There is no controversy to teach because mainstream science recognizes that evolution is the best explanation for the great variety of life on Earth; it is a fact. This is why the National Academy of Sciences and the American Association for the Advancement of Science (publishers of the journal *Science*) have publicly expressed their opposition to teaching ID in public schools. The idea of a controversy is simply a clever fabrication designed by ID publicists and media hounds.

Teaching ID in schools as if it were a genuine scientific alternative to the theory of evolution will only confuse students about how science really works. And further, it will put those students who might pursue scientific careers at a serious educational disadvantage.

We've seen that ID fails for a number of reasons: It simply repackages the discredited argument from design. It falsely claims that biological complexity can't be explained without a designer. It wrongly states that using natural scientific methods requires atheism. And its adherents pretend there's a scientific controversy about evolution when there is none. Ultimately, ID is not genuine science because it offers no scientific evidence in its support and has no coherent scientific methods or mechanisms.[34]

## NOTES

1. W. Paley, *Natural Theology* (New York: American Tract Society, n.d.), p. 1.

2. C. Darwin, *On the Origin of Species by Means of Natural Selection*, 1st ed. (London: Murray, 1859), p. 186, http://pages.britishlibrary.net/charles.darwin/texts/origin1859/origin_fm.html (accessed November 18, 2005).

3. For a thorough explanation of how eyes have evolved in the Animal Kingdom, see R. Dawkins, *Climbing Mount Improbable* (New York: Norton, 1996), chap. 5.

4. C. B. Thaxton, W. L. Bradley, and R. L. Olson, *The Mystery of Life's Origin: Reassessing Current Theories* (New York: Philosophical Library, 1984).

5. Actually, the phrase "intelligent design" was used by the Irishman John Tyndall in an 1874 address to the British Association, where he described the views of the first-century BCE Roman philosopher Lucretius, and the rejection by Lucretius of intelligent design as an explanation for the organization of matter. See J. Tyndall, *Address Delivered before the British Association Assembled at Belfast, with Additions* (London: Longmans & Green, 1874), p. 8.

6. M. J. Behe, *Darwin's Black Box: The Biochemical Challenge to Evolution* (New York: Free Press, 1996), p. 39.

7. Ibid., p. 39.

8. Ibid., p. 193.

9. H. A. Orr, "Darwin v. Intelligent Design (again): The Latest Attack on Evolution Is Cleverly Argued, Biologically Informed—and Wrong," *Boston Review* (December 1996/January 1997), http://www.bostonreview.net/BR21.6/orr.html (accessed November 20, 2005).

10. D. J. Futuyma, "Miracles and Molecules," *Boston Review* (February/March 1997), http://www.bostonreview.net/br22.1/futuyma.html (accessed November 20, 2005).

11. Ibid.

12. Orr, "Darwin v. Intelligent Design (again)."

13. For a good account of the evidence for the gradual evolution of bacterial flagella, see D. Ussery, "Darwin's Transparent Box: The Biochemical Evidence for Evolution," in *Why Intelligent Design Fails: A Scientific Critique of the New Creationism*, ed. M. Young and T. Edis (New Brunswick, NJ: Rutgers University Press, 2005), pp. 48–57.

14. M. Young and T. Edis, *Why Intelligent Design Fails*, p. 3.

15. Behe acknowledged his belief that the intelligent designer is God at the Religious Roots of Liberty conference, sponsored by the Philadelphia Society, April 25–27, 1997.

16. K. R. Miller, review of *Darwin's Black Box: The Biochemical Challenge to Evolution*, by M. J. Behe, *Creation/Evolution* 16, no. 2 (1996): 36.

17. P. Davis and D. H. Kenyon, *Of Pandas and People* (Dallas, TX: Haughton, 1993).

18. W. Frair and P. Davis, *A Case for Creation*, 3rd ed. (Chicago: Moody, 1983), p. 94. Quoted in R. T. Pennock, *Tower of Babel: The Evidence against the New Creationism* (Cambridge, MA: Bradford Books/MIT Press, 1999), p. 192.

19. St. Ignatius of Loyola, *The Spiritual Exercises of St. Ignatius of Loyola*, translated from the autograph by Father Elder Mullan, S. J. (New York: Kennedy & Sons, 1914), p. 75, http://www.ccel.org/ccel/ignatius/exercises.pdf (accessed January 10, 2006).

20. R. T. Pennock, "The Wizard of ID: Reply to Dembski," in *Intelligent Design Creationism and Its Critics: Philosophical, Theological, and Scientific Perspectives*, ed. R. T. Pennock (Cambridge, MA: MIT Press, 2001), p. 647.

21. W. A. Dembski, "What Every Theologian Should Know about

Creation, Evolution, and Design," *Center for Interdisciplinary Studies Transactions* 3, no. 2 (1995): 3. Quoted in R. T. Pennock, "The Wizard of ID," p. 648. For more on Dembski's inconsistent position, see R. T. Pennock, *Intelligent Design Creationism and Its Critics*, chap. 30.

22. B. Forrest and P. R. Gross, *Creationism's Trojan Horse: The Wedge of Intelligent Design* (Oxford: Oxford University Press, 2004).

23. *Philosophical naturalism* is also called *metaphysical naturalism*, or *ontological naturalism*.

24. For a clear account of distinct varieties of naturalism and how Intelligent Design Creationists ignore these distinctions, see R. T. Pennock, *Tower of Babel*, chap. 4.

25. Some philosophers argue that if science and reason can't tell us whether God exists, then belief in God is wholly irrational in the way that belief in Zeus or Thor is irrational.

26. W. A. Dembski, *No Free Lunch: Why Specified Complexity Cannot Be Purchased without Intelligence* (New York: Rowman & Littlefield, 2002), pp. 347–53. See also M. J. Behe, *Darwin's Black Box*, p. 251.

27. Behe testified in court on October 18, 2005, in support of the Dover, Pennsylvania school board's efforts to teach ID (*Kitzmiller et al. v. Dover Area School District*). Under cross-examination Behe admitted that his definition of "scientific theory" was so broad that it would include astrology. P. E. Johnson, *The Wedge of Truth: Splitting the Foundations of Naturalism* (Downers Grove, IL: InterVarsity, 2000), p. 155.

28. Johnson wrongly assumes that evolutionary scientists must support *philosophical naturalism* as opposed to *methodological naturalism*, and that their position involves by default the denial of the existence of God. For more on how Johnson ignores important conceptual distinctions, see note 24.

29. Ibid., p. 151.

30. Ibid., p. 163.

31. For an analysis and critique of Johnson's "theistic science," see R. T. Pennock, "The Prospects for a 'Theistic Science,'" *Perspectives on Science and Christian Faith* 50 (September 1998): 205–209.

32. "Intelligent Design Not Science, Says Vatican Newspaper Article," *Catholic News Service* (January 17, 2006), http://www.catholicnews.com/data/stories/cns/0600273.htm (accessed January 18, 2006).

33. There are a number of philosophical problems with regard to the-

ological claims, and many philosophers have found the solutions offered by theologians to be unsatisfactory. Some of these claims include the existence of, and explanations for, God, angels, miracles, prophesies, and immortality. A problem specific to evolution concerns why a morally good creator would permit natural selection—a process rife with death and suffering—to be a necessary mechanism in the appearance of the diverse life-forms on Earth, including humans. Darwin expressed his concern about this in a letter to American botanist Asa Gray, writing, "I cannot persuade myself that a beneficent and omnipotent God would have designedly created the Ichneumonidae [a kind of wasp] with the express intention of their feeding within the living bodies of Caterpillars, or that a cat should play with mice." And in a letter to his friend Joseph Hooker, Darwin writes, "What a book a devil's chaplain might write on the clumsy, wasteful, blundering, low, and horribly cruel works of nature!" For a letter to Gray, see F. Darwin, ed., *The Life and Letters of Charles Darwin* (New York: Appleton, 1905), vol. 2, chap. 2, p. 105, http://pages.britishlibrary.net/charles.darwin/texts/letters/letters2_02.html (accessed September 14, 2006). For letter to Hooker, see F. Darwin and A. C. Seward, eds., *More Letters of Charles Darwin* (London: John Murray, 1903), vol. 1, chap. 2, p. 94, http://pages.britishlibrary.net/charles.darwin/texts/more_letters/mletters1_02.html (accessed September 14, 2006).

34. Under cross-examination in *Kitzmiller et al. v. Dover Area School District* (on October 18, 2005), Behe claimed that natural selection alone could not account for life's complexity, but he was unable to identify any mechanism that could account for complexity.

## Myth Ten

# Evolution Is Immoral

In January 1860, just a few months after the publication of *On the Origin of Species*, Darwin sat down to compose a letter to his geologist friend Charles Lyell, stating, "I have received, in a Manchester newspaper, rather a good squib, showing that I have proved 'might is right,' and therefore that Napoleon is right, and every cheating tradesman is also right."[1] Darwin was rightfully shocked to discover that his views on evolution had been interpreted to support the view that "might is right." Nowhere in *On the Origin* does he imply any such thing. Rather, he avoided discussing human evolution and morality in that book.

So how did evolution get tied up with the view that might makes right? In 1851 the Englishman Herbert Spencer (1820–1903) published *Social Statics*, a book in which he applied the idea of evolution to the natural changes in the institutions and the organization of societies. Spencer's take on evolution—as it was understood at the time—was not Darwinian, and it couldn't have been since Darwin's *On the Origin* didn't appear until nine years later.[2] Spencer's view was Lamarckian. He believed that human evolution—particularly the evolution of societies—had a natural goal, and that this goal involved upward progress to higher and better stages. This idea of progress was so wedded to evolution that even many late nineteenth-century anthropologists mistakenly believed that all societies follow a natural progression upward from "savagery" through "barbarism" to civilization. It's no surprise that European societies were seen as the highest form of culture and civilization. For Spencer, European civilization still had room to progress, and one of the keys to further evolution was limiting the role of government in the economic sphere. Governments, Spencer believed, tend to hold back this natural evolution when they interfere in the affairs of the business class and when they provide economic and social services to the lower classes.

Spencer was not against individuals or private charities helping the poor, but his views on extreme free-market economics ruled out government aid. Spencer was a careful thinker, and it would be a mistake to write him off as someone who simply thought that "the survival of the fittest"—a phrase he coined—was the basis for morality. Yet, in perhaps one of his less careful moments, he implied that those individuals who do survive are those who ought to survive:

> Nature demands that every being shall be self-sufficing. All that are not so, nature is perpetually withdrawing by death. . . . If they are sufficiently complete to live, they do live, and it is well they should live. If they are not sufficiently complete to live, they die, and it is best they should die.[3]

Spencer's emphasis on self-reliant individualism in the above passage and his catchy phrase "the survival of the fittest" suggested to many that nature was a realm of competition that could provide a framework for how societies should be structured and, ultimately, for how we ought to live.

In contrast, the Russian prince turned anarchist, Peter Kropotkin (1842–1921), had a different take on human evolution and society. For Kropotkin, individual self-sufficiency and competition were not so important. Rather, he saw a natural tendency toward sympathy and cooperation in humans (and many animals), which he called *mutual aid*. According to Kropotkin, species that cooperated most (and competed least) among themselves would inevitably be the most successful, while "the unsociable species, on the contrary, are doomed to decay."[4]

Some thought that social progress, and perhaps even moral progress, could be had by emphasizing certain natural tendencies in humans. The big question was whether we should rely on competitive or cooperative tendencies for guidance.

A third view was proposed by Thomas Henry Huxley (1825–1895)—known as "Darwin's Bulldog" for his vigorous support of Darwin's theory. He was vehemently against looking to nature for social or moral guidance. Huxley saw the realm of nature—what he called the *cosmic process*—as a bloody and brutal arena, which was not only unsuitable for providing moral guidance, but dangerously so, writing,

> Let us understand, once and for all, that the ethical process of society depends, not on imitating the cosmic process, still less in running away from it, but in combating it.[5]

For Huxley, the competitive, cutthroat processes of nature and survival were something to be resisted by civilized people.

The views of Spencer, Kropotkin, and Huxley can be interpreted as supporting three completely different positions on how nature and evolution should inform (if at all) our social and moral practices. That analyzing the workings of nature could spawn three such opposing views should have given pause. Yet the idea of finding a basis for morality in nature continued to have great appeal for the general public for some time.

## The Business of Social Darwinism

The business tycoon John D. Rockefeller (1839–1937) built an oil monopoly, Standard Oil, by outcompeting and buying up other companies. Rockefeller supported an extreme form of free-market capitalism, arguing strenuously against the government's attempt to break up his empire.[6] He saw the workings of the free market as ordained by both nature and God. In a famous Sunday-school address about the merits of economic competition, Rockefeller said:

> The growth of a large business is merely a survival of the fittest. . . . The American Beauty rose can be produced in the splendor and fragrance which brings cheer to its beholder only by sacrificing the early buds which grow up around it. This is not an evil tendency in business. It is merely the working-out of a law of nature and a law of God.[7]

Spencer's view of natural progress, and Darwin's observation that nature weeded out the less fit, became intertwined in the popular culture of the late nineteenth and early twentieth century, in what came to be known as *social Darwinism*. It seemed obvious and natural to Rockefeller and others, that, if you wanted to form a more advanced and better functioning society (particularly in economics), then the best thing to do was to let the fittest control the direction of society's "evolution." It wasn't the strongest or those best at cooperation who were considered the fittest, but those who came out on top in competition in business.

Another industrialist, Andrew Carnegie (1835–1919), was also persuaded by how the "law of nature" might justify the concentration of wealth in the hands of a few when he argued, "While the law may sometimes be hard for the individual, it is best for the race, because it ensures the survival of the fittest in every department."[8] Social Darwinism, with its emphasis on competition, seems to be a rather self-serving justification for the rich to get richer, and the powerful to become more powerful, at the expense of those below. If social Darwinism and its implications for wealth and power seem morally troublesome, it's nothing compared to the eugenics movements of the first half of the twentieth century.

## Taking Stock

Where social Darwinists aimed to prevent interference with the natural progression of society, as they saw it, supporters of eugenics thought it better to give natural selection a helping hand. One way was to encourage the "best" people to reproduce, which would ensure that future generations would be more fit, biologically. A more troubling approach was to prevent the less fit from reproducing.[9]

In 1907 the state of Indiana passed a law allowing "defectives" to be sterilized. The idea was to weed out the "bad stock," particularly the criminally insane, and the so-called idiots, imbeciles, and the feebleminded, so that they wouldn't pass on their traits to the next generation. By 1933 twenty-nine other states had similar laws on the books. Between 1907 and 1974 it's estimated that hundreds of thousands of "defectives" were sterilized in the United States, often against their will.[10] The fear of race-mixing was also troubling to many, as was the worry about the high birthrates among immigrants of "inferior races." President Theodore Roosevelt was so troubled by the expansion and encroachment of these "undesirables" that he warned of a "war of the cradle" being fought between the "superior" and the "inferior" social classes.[11]

Throughout the 1920s and 1930s, sterilization laws were enacted in Denmark, Finland, Sweden, Norway, and Estonia, many of which were modeled after American laws.[12] But social Darwinism was catapulted to extremes by the policies of the German Nazi Party. The Nazis attempted to purify the blood of the German people through sterilization, and eventually through "euthanasia," or so-called mercy killing. Between 1933 and 1945, about 360,000 Germans were sterilized.[13] The Nazis also had a cruel policy of "euthanizing" the handicapped. But they finally went to horrific lengths by exterminating over six million Jews and hundreds of thousands of others who they perceived as members of inferior races.[14]

Ultimately, the attempt to morally justify the policies of social Darwinism and eugenics involved misapplying our understanding of evolution, either by assuming that unfettered competition was natural, that race-mixing was biologically harmful, or that there is such a thing as inferior races. Another grave mistake was to assume that just because something is natural—perhaps even a product of evolution—that it must be good.

## Is Natural Good?

The idea that not only do the fittest tend to survive, but also that they *ought to* survive runs into a quagmire of moral confusion associated with deriving moral values from properties of nature. The problem is that just because something is natural, it doesn't follow that it's good, assuming we even have a clear understanding of the word *natural*.[15] Natural things may or may not be good. We can't just assume they are. Consider the Ebola virus. It's certainly natural, but is it good? The disease caused by the virus, Ebola hemorrhagic fever, has no known cure. It kills 50–90 percent of people infected, and its symptoms include vomiting, diarrhea, liver and kidney impairment, and sometimes internal and external bleeding.[16] Some victims end up bleeding to death out of every orifice of their body. It's a horrible way to die. As natural as it is, the Ebola virus is certainly not good for those infected with it.

On the other hand, antibiotics might be considered unnatural—thus, not good—since they're created by humans in laboratories. But most of us would consider antibiotics a good thing because they fight potentially deadly infections, which are themselves quite natural. While some natural things are better than synthetic alternatives, it would be a mistake to assume that just because something is natural it must be good. Consider also that death itself is quite natural, although we don't usually consider it a good thing.

It's quite common to connect *natural* with *good* in everyday language. Many of us consider natural food better than "unnatural" food, such as real cheese as opposed to neon-orange "cheez" sprayed out of a can. The word *unnatural* has negative connotations in common usage, and *natural* has positive ones. And this linking of natural with good probably contributed to the easy acceptance of social Darwinism, which seems the case with Rockefeller's attempt to justify extreme free-market competition by comparing it to cultivating roses.

We can ask not just whether organisms found in nature are good for us, but whether certain human behaviors that are considered natural are also good. Displays of aggressive behavior seem to be quite natural among many animals. Humans can also behave aggressively, and this behavior with its accompanying emotions is surely an evolutionary inheritance, what we sometimes call our *animal nature.* We can confidently say that aggressive behavior is natural for humans, unlike, say, eating with a fork or chopsticks, which are cultural options. But if acting aggressively is natural, we can ask if it's a morally good thing. There may be cases when aggression is good or at least morally justifiable, as when a mother acts with hostility toward someone who poses a threat to her children. But all of us have been in a bad mood or have just had a bad day, and we've all lashed out at someone for no good reason. In such cases acting aggressively would not be morally good behavior.

There are a host of natural or animal impulses and behaviors, such as anger, jealousy, fear, and lust, that can affect how we treat others, and it's obvious that acting on these impulses is not always a good thing morally.

If it could be shown—however unlikely—that rape and murder were natural human behaviors that are part of our evolutionary inheritance, we can still ask whether these behaviors are morally good. It should be clear—to most of us, at least—that these sorts of behaviors are morally wrong. Therefore, it would be incumbent on us to combat them by teaching children from an early age to resist the urge to act on such impulses.

Ultimately, we must reject the assumption that just because something is natural, or even a result of biological evolution, that it must be good and worthy of imitation. Thanks to culture, humans are rather flexible and adaptive, and we can learn to resist many unethical behavioral tendencies that may have an evolutionary basis.

## Divine Commands

Religions are powerful cultural institutions that provide structure and meaning in many people's lives, often by providing moral rules for how we ought to behave. And many fundamentalist Christians hold to the view that you cannot be a moral person if you accept evolution. This is because they believe that God is the source of morality and that accepting evolution requires you to be an atheist. As we saw in "Myth Nine: Intelligent Design Is Science," the belief that science in general, and that evolution in particular, requires atheism rests on a mistake. But the idea that God is the source of morality and that only those who believe in God can be genuinely moral people is a pervasive, powerful, but ultimately mistaken idea.

On this view, God is the sole author of morality, in that he makes right actions right and wrong actions wrong by his commands. Acting morally, then, is simply a matter of following God's commands. Behaviors such as lying, stealing, and killing are morally wrong *because* God forbids them. Likewise, actions such as telling the truth and keeping one's promises are morally right *because* God commands them. This view of the nature of morality is called the *divine command theory*.

Over 2,300 years ago, the Greek philosopher Plato (428–348 BCE), in his dialogue *Euthyphro*, convincingly showed why the divine command theory creates problems.[17] Supporters of the divine command theory claim that an action is morally wrong *because* God says it's wrong. In other words, behaviors such as killing and stealing are morally wrong because God *makes them wrong* by his commands. If God did not forbid us to kill and steal, then killing and stealing would not be morally wrong. Likewise, morally right behaviors, such as being honest and keeping one's promises, are right only because God *says* they're right. The problem with this view is that whether an action is wrong (or right) becomes completely arbitrary. God could have said that killing and stealing are morally right, and then those actions would be morally right. You might object that God would never say that killing and stealing are morally right. But why not? According to the divine command theory, killing and stealing were not morally wrong *before* God made them wrong by his commands. In other words, God did not first recognize the wrongness of killing and stealing, and then command us not to kill and steal. How could God recognize that killing and stealing were wrong if those actions weren't wrong to begin with? There would be nothing to recognize about those actions that make them wrong. Remember, according to the divine command theory, it is God's commands that *make* actions wrong, and they were not wrong before God forbade them. So, it seems that there can be no reason for God to decide to make killing and stealing wrong, and thus his decisions are arbitrary.

The way around this problem is to argue that God's commands don't *make* certain actions morally wrong. Rather, God *sees* or *recog-*

*nizes* that actions such as killing and stealing are wrong, and that's why he forbids them. By making this move, we get around the problem of God's commands being arbitrary, but in the process we're forced to reject the divine command theory. We're left with morality being independent of God in the same way that arithmetic and logic are independent of God. God does not make it true that 2 + 2 = 4. Instead, he recognizes that it's true.[18] Similarly, God does not make killing and stealing wrong. Instead, he recognizes the reasons that make killing and stealing wrong, and he forbids us to kill and steal because of these compelling reasons.

But notice that, because morality is independent of God, we can recognize the same reasons for not killing and not stealing that God recognizes, although our thinking is certainly slower. We recognize the irreversible harm caused by killing as a reason not to kill, and we recognize the unfairness of stealing as a reason not to steal. Because morality is independent of God, both the believer and the nonbeliever are in the same boat when it comes to making moral choices. So we can see that one does not have to believe in God in order to be a genuinely moral person. But if morality is independent of God, where does it come from?

## The Evolution of Moral Emotions

It's unlikely that understanding human evolution will ever provide us with specific moral rules or principles for how we ought to treat one another. And perhaps Huxley was right to warn against looking to nature for moral guidance. But the realm of morality includes more than our ability to reason about principles and how we ought to act. It also includes an emotional component. We have certain moral feelings, such as empathy, shame, gratitude, and indignation, that more than likely have an evolutionary history. Understanding the origins of these feelings might give us some insight into the biological roots of human morality.

Since the fossil remains of our ancient ancestors can't tell us much about the evolution of our moral emotions, perhaps taking a look at the behavior of other animals might give us some clues. Consider that all mammals are born dependent on their mothers, and that care for the young is essential for their survival. If mothers did not feel emotionally connected to their infants, those infants would probably die. The evolution of maternal care of offspring in mammals can be explained fairly easily: those mothers who didn't care for their young would end up with dead offspring, whereas the offspring of mothers who cared for their young tended to survive, with the mothers' genes for maternal care passing on to their infants. Human mothers also care for their young. And while our attitudes about the importance of caring for children are certainly passed on through culture, language, and learning, it would be silly to think that being mammals has nothing at all to do with our caring for our young.

When we observe chimpanzees (our closest, living, nonhuman relatives) we don't find a full-blown morality, but we do find what might be called *premoral emotions*. Chimpanzees are rather intelligent, social animals who live in communities of about twenty to sixty in the wild. Members of the same group frequently engage in a kind of tit-for-tat sharing known as *reciprocal altruism*. This typically involves sharing food (especially meat) and exchanging grooming services. Like humans, chimps get along well with some of their fellows better than others, and keeping track of just where they stand in terms of the closeness or distance of their relationships requires a good memory. In fact, chimpanzees seem to remember which of their fellows have done them favors in the past and which haven't. While observing chimpanzees, the ethologist Frans de Waal found that

> . . . if A shared a lot with B, B generally shared a lot with A, and if A shared little with C, C also shared little with A. [And] . . . grooming affected subsequent sharing: A's chances of getting food from B improved if A had groomed B earlier that day.[19]

Chimps can behave in ways that are eerily familiar. They'll commonly console those who have just been in a fight—particularly the losers—by hugging them, grooming them, or patting them on the back.[20] Lower-ranking chimps may join together to stand up against a dominant, bullying chimp. And a dominant chimp may also retaliate later against one of the lower-ranking chimps who ganged up on him, particularly when the lower-ranking chimp's buddies are not around.[21] And perhaps most strikingly, chimps have also been known to show not only tolerance, but also care and protection toward handicapped members of their group.[22]

Of course chimpanzees are not humans, but it's not a stretch to think that some of our moral feelings or sentiments are the products of evolution, and that living in highly social communities, as our early human ancestors did, selected for many of the moral sentiments we feel today. Perhaps our feelings of empathy for the suffering of others, gratitude when treated kindly, or indignation at being treated meanly have their roots in our evolutionary past. These feelings are not themselves a full-blown morality, but more like an emotional foundation that is necessary in order to have a morality at all. Darwin thought it highly probable that

> [a]ny animal whatever, endowed with well-marked social instincts, the parental and filial affections being here included, would inevitably acquire a moral sense or conscience, as soon as its intellectual powers had become as well, or nearly as well developed, as in man.[23]

If we consider culture, language, and reason as part of these intellectual powers, then Darwin's view seems quite plausible; we're a social species who rely on culture, language, and reason. That's our nature. For over one hundred thousand years, anatomically modern humans lived as hunter-gatherers in groups averaging about 150 individuals, and it's only in the last ten thousand years—with the development of agriculture—that humans commonly began to live in much larger groups.[24] Living as hunter-gatherers requires cooperation and conformity. Strong cultural and social pressures would have been felt by those who did not conform, and by those who endangered the harmony or the very existence of the group.

Since morality is fundamentally about our relations with others, it seems reasonable to think that certain universal human moral emotions evolved under the pressure of maintaining relatively harmonious relationships while living in highly social groups. Although different cultures may vary in what they perceive as morally acceptable or reprehensible behavior, moral feelings such as empathy, shame, indignation, and gratitude are universal.[25] Similarly, all societies must have prohibitions against murder, lying, and stealing if they are to exist at all, although what counts as justified exceptions to these prohibitions may differ among societies.[26]

## A Modest Light

Ultimately, it would be a mistake to seek moral guidance from our understanding of human evolution. Social Darwinism fails because it selects certain kinds of human behaviors—typically competitive behaviors—calls them natural in an "evolutionary sense," and then assumes that because they're natural they must be morally good. But we can see that natural is not always good.

Accepting the truth of evolution does not lead to immorality, as some religious fundamentalists believe. Since morality is independent of God, as Plato showed, all humans are in the same boat when it comes to weighing the moral reasons for and against certain actions.

Finally, we can be fairly confident that many of our moral emotions have a biological basis and an evolutionary history, which must have served our species well in the past as our ancestors navigated the often turbulent waters of social interactions. By examining these emotions we can shed some light on the complexity of our social nature and on how the intricacies of our social relationships evolved over time.

## Notes

1. F. Darwin, ed., *The Life and Letters of Charles Darwin* (New York: Appleton, 1905), vol. 2, chap. 2, pp. 56–57, http://pages.britishlibrary.net/charles.darwin/texts/letters/letters1_fm.html (accessed December 5, 2005).

2. Even after Darwin's *On the Origin of Species* was published, Spencer still maintained a Lamarckian view of evolution.

3. H. Spencer, *Social Statics: The Conditions Essential to Happiness Specified, and the First of Them Developed* (New York: Robert Schalkenbach Foundation, 1970), pp. 339–40.

4. P. Kropotkin, *Mutual Aid: A Factor of Evolution* (Boston: Extending Horizons Books, 1955), p. 292.

5. T. H. Huxley, *Evolution and Ethics and Other Essays* (New York: Greenwood, 1968), p. 83.

6. In 1911 Standard Oil was ordered broken up into thirty-four separate, smaller companies when the Supreme Court declared it a monopoly under the Sherman Antitrust Act.

7. R. Hofstadter, *Social Darwinism in American Thought* (Boston: Beacon, 1983), p. 45.

8. Ibid., p. 46.

9. N. Levy, *What Makes Us Moral? Crossing the Boundaries of Biology* (Oxford: Oneworld, 2004), pp. 14–15.

10. M. Hawkins, *Social Darwinism in European and American Thought, 1860–1945: Nature as Model and Nature as Threat* (Cambridge: Cambridge University Press, 1997), p. 242.

11. A. Buchanan et al., *From Chance to Choice: Genetics and Justice* (Cambridge: Cambridge University Press, 2000), p. 38.

12. M. E. Kopp, "Eugenic Sterilization Laws in Europe," *American Journal of Obstetrics and Gynecology* 34 (September 1937): 499.

13. M. Hawkins, *Social Darwinism in European and American Thought, 1860–1945*, p. 279.

14. For a thorough history of the German Nazi Party and its policies,

see W. L. Shirer, *The Rise and Fall of the Third Reich: A History of Nazi Germany* (New York: Simon & Schuster, 1960).

15. The view presented here—that it doesn't follow that because something is natural that it must be good—is not quite the same as the philosopher G. E. Moore's *naturalistic fallacy*, which analyzes the concepts *natural* and *good*, and argues that they cannot mean the same thing. See G. E. Moore, *Principia Ethica* (Cambridge: Cambridge University Press, 1903).

16. World Health Organization, *Ebola Haemorrhagic Fever Fact Sheet*, http://www.who.int/mediacentre/factsheets/fs103/en (accessed December 12, 2005).

17. In the dialogue *Euthyphro*, Plato has Socrates arguing against the view that it's the gods (rather than God) who make right actions right. It might seem that this argument doesn't apply to monotheism. But once Euthyphro and Socrates agree to limit the argument only to those actions that the gods unanimously agree upon to make right, then the argument is easily applied to monotheism.

18. It might be thought that if God created the universe, he also created mathematical and logical truths and is not bound by these truths. But St. Thomas Aquinas (1225?–1274) seemed to think that God cannot do the logically impossible when he wrote, "Whatever implies contradiction does not come within the scope of divine omnipotence, because it cannot have the aspect of possibility." And the Christian author C. S. Lewis seemed to agree, writing, "His Omnipotence means power to do all that is intrinsically possible, not to do the intrinsically impossible." For Aquinas quotation, see Thomas Aquinas, *The Summa Theologica* (I, question 25, article 3) (New York: Benziger Bros., 1947), trans. Fathers of the English Dominican Province, http://www.ccel.org/a/aquinas/summa/FP/FP025.html#FPQ25A3THEP1 (accessed September 14, 2006). For Lewis quotation, see C. S. Lewis, *The Problem of Pain* (San Francisco: Harper, 2001), p. 18.

19. F. B. M. de Waal, *Good Natured: The Origins of Right and Wrong in Humans and Other Animals* (Cambridge, MA: Harvard University Press, 1996), p. 153.

20. Ibid., pp. 60–61.

21. Ibid., pp. 157–58.

22. Ibid., pp. 44–53.

23. C. Darwin, *The Descent of Man and Selection in Relation to Sex*, 2nd

ed., revised and augmented (London: Murray, 1882), p. 98, http://pages.britishlibrary.net/charles.darwin/texts/descent/descent_front.html (accessed December 18, 2005).

24. By one hundred thousand years ago, humans were anatomically modern and were behaving in ways that were somewhat modern. See R. G. Klein, *The Human Career* (Chicago: University of Chicago Press, 1999). Anthropologist Robin Dunbar has calculated that the average hunter-gatherer group size is 149.8 individuals (although such groups probably split up into smaller bands for months at a time). See R. I. M. Dunbar, "Neocortex Size as a Constraint on Group Size in Primates," *Journal of Human Evolution* 20 (1992): 469–93. Concerning the origins of agriculture and its effect on population density, see R. J. Wenke, *Patterns in Prehistory: Humankind's First Three Million Years* (New York: Oxford University Press, 1999), pp. 268–329.

25. M. Shermer, *The Science of Good and Evil: Why People Cheat, Gossip, Care, Share, and Follow the Golden Rule* (New York: Holt, 2004), pp. 285–92.

26. J. Rachels, *The Elements of Moral Philosophy*, 2nd ed. (New York: McGraw-Hill, 1993), pp. 25–26.

# Afterword

In this book we've seen that evolution isn't "just a theory," but the undisputed, scientifically supported fact at the foundation of the life sciences. We've also seen that evolution doesn't climb a ladder of progress toward human beings, and that it doesn't have a conscious plan to follow a particular direction or to balance nature for the needs of living organisms. Still, order, patterns, and progress do occur, but, fascinatingly, these result from the undesigned effects of the observable facts of evolution: replication, variation, and selection.

We've also seen that "survival of the fittest" doesn't mean that only the biggest and strongest survive through ruthless competition, whether among humans or among other animals. We've also seen that we cannot justifiably use this phrase to support the immoral practices of economic slavery, involuntary euthanasia, or ethnic cleansing.

We've observed that the creationist arguments against evolution—such as that Earth is very young—are simply wrong. The facts of geology have proven that Earth is billions, not thousands, of years old. We've also observed how a new form of creationism, intelligent design, isn't a genuine scientific theory after all, but the same old creationism dressed up in a new suit. We've seen that the evolutionary transitions between species are better seen as shades or grades rather than discrete links. And we've seen that even though humans have language and culture and are extremely intelligent, the facts of biology show that we're still animals, and more specifically, we're primates.

In the end, we must recognize that the processes of evolution explain not only the origins of bacteria, viruses, plants, and other animals, but also the origin of our own species. To better understand the living world and our relationship to it, we need to battle the antiquated myths that distort our understanding of how evolution really works. And we need to battle the deliberately misleading arguments of religious ideologists who wish to replace the teaching of evolution with biblical literalism or theologically inspired "science." We hope that this book has equipped many for these battles.

Cameron McPherson Smith
Charles Sullivan

# Bibliography

Alvarez, W., and F. Asaro. "An Extraterrestrial Impact (Accumulating Evidence Suggests an Asteroid or Comet Caused the Cretaceous Extinction)." *Scientific American* 263, no. 4 (October 1990): 78–84.

Aquinas, Thomas, St. *The Summa Theologica* (I, question 25, article 3). New York: Benziger Bros., 1947. Translated by Fathers of the English Dominican Province. http://www.ccel.org/a/aquinas/summa/FP/FP025.html#FPQ25A3THEP1 (accessed September 14, 2006).

Asfaw, B., T. White, O. Lovejoy, B. Latimer, S. Simpson, and G. Suwa. "*Australopithecus garhi*: A New Species of Early Hominid from Ethiopia." *Science* 284, no. 5414 (1999): 629–35.

"Asteroid and Comet Impact Hazards." http://impact.arc.nasa.gov/ (accessed January 22, 2005).

Badkhen, A. "Anti-Evolution Teachings Gain Foothold in U.S. Schools." *San Francisco Chronicle*, November 30, 2004.

Bar-Yosef, O. "Eat What Is There: Hunting and Gathering in the World of Neanderthals and Their Neighbours." *International Journal of Osteoarchaeology* 14, nos. 3–4 (2004): 333–42.

Becker, L., R. J. Poreda, A. R. Basu, K. O. Pope, T. M. Harrison, C. Nicholson, and R. Iasky. "Bedout: A Possible End-Permian Impact Crater Offshore of Northwestern Australia." *Science* 304, no. 5676 (2004): 1469–76.

Behe, M. J. *Darwin's Black Box: The Biochemical Challenge to Evolution*. New York: Free Press, 1996.

Benton, M. J. *When Life Nearly Died: The Greatest Mass Extinction of All Time.* London: Thames & Hudson, 2003.

Benton, M. J., and P. N. Pearson. "Speciation in the Fossil Record." *Trends in Ecology and Evolution* 16, no. 7 (2001): 405–11.

Berra, T. M. *Evolution and the Myth of Creationism: A Basic Guide to the Facts in the Evolution Debate.* Stanford, CA: Stanford University Press, 1990.

Black, R. "'New Mammal' Seen in Borneo Woods." http://news.bbc.co.uk/1/hi/sci/tech/4501152.stm (accessed December 6, 2005).

Blackwell, L. R., and F. D'Errico. "Evidence of Termite Foraging by Swartkrans Early Hominids." *Proceedings of the National Academy of Sciences USA* 98, no. 4 (2001): 1358–63.

"BodyBurden: The Pollution in Newborns." http://www.ewg.org/reports/bodyburden2/ (accessed December 22, 2005).

Bos, L., ed. *Plant Viruses: Unique and Intriguing Pathogens: A Textbook of Plant Virology.* Leiden, Netherlands: Backhuys, 1999.

Boyle, K. V. "Reconstructing Middle Palaeolithic Subsistence Strategies in the South of France." *International Journal of Osteoarchaeology* 10, no. 5 (2000): 336–56.

Britt, A. "DNA Damage and Repair in Plants." *Annual Review of Plant Physiology and Plant Molecular Biology* 47 (1996): 75–100.

Broom, R. *Finding the Missing Link.* London: Watts, 1950.

Buchanan, A., N. Daniels, D. Winkler, and D. Brock. *From Chance to Choice: Genetics and Justice.* Cambridge: Cambridge University Press, 2000.

Cain, M. L., B. G. Milligan, and A. E. Strand. "Long-Distance Seed Dispersal in Plant Populations." *American Journal of Botany* 87, no. 9 (2000): 1217–27.

Chen, J. Y., J. Dzik, G. D. Edgecombe, L. Ramskold, and G. Q. Zhou. "A Possible Early Cambrian Chordate." *Nature* 377 (1995): 720–22.

Cifelli, R. L., and J. G. Eaton. "Marsupial from the Earliest Late Cretaceous of Western US." *Nature* 325 (1987): 520–22.

Clutton-Brock, J. *A Natural History of Domesticated Mammals.* Cambridge: Cambridge University Press, 1999.

Cooper, A., C. Mourer-Chauvire, G. K. Chambers, A. von Haeseler, A. C. Wilson, and S. Paabo. "Independent Origins of New Zealand Moas and Kiwis." *Proceedings of the National Academy of Sciences USA* 89, no. 18 (1992): 8741–44.

Cowen, R. *History of Life.* 4th ed. Malden, MA: Blackwell Science, 2005.

Cuddington, K. "The 'Balance of Nature' Metaphor and Equilibrium in Population Ecology." *Biology and Philosophy* 16 (2001): 463–79.

Dalrymple, G. B. *The Age of the Earth.* Stanford, CA: Stanford University Press, 1991.

———. "How Old Is the Earth? A Reply to 'Scientific' Creationism." In vol. 1, pt. 3, *Evolutionists Confront Creationism: Proceedings of the 63rd Annual Meeting of the Pacific Division, American Association for the Advancement of Science*, edited by F. Awbrey and W. Thwaites, 66–131. AAAS, 1984.

Darwin, C. *The Descent of Man and Selection in Relation to Sex.* 2nd ed., revised and augmented. London: Murray, 1882. http://pages.britishlibrary.net/charles.darwin/texts/descent/descent_front.html (accessed June 14, 2005 and December 18, 2005).

———. *On the Origin of Species by Means of Natural Selection.* 1st ed. London: Murray, 1859. http://pages.britishlibrary.net/charles.darwin/texts/origin1859/origin_fm.html (accessed November 18, 2005).

———. *On the Origin of Species by Means of Natural Selection.* 6th ed. London: Murray, 1872. http://pages.britishlibrary.net/charles.darwin/texts/origin_6th/origin 6th _fm.html (accessed June 14, 2005).

———. *The Variation of Animals and Plants under Domestication.* 2nd ed. Vol. 2. New York: Appleton, 1883. http://pages.britishlibrary.net/charles.darwin/texts/variation/variation_fm1.html (accessed June 20, 2005).

Darwin, F., ed. *The Life and Letters of Charles Darwin.* New York: Appleton, 1905. http://pages.britishlibrary.net/charles.darwin/texts/letters/letters1_fm.html (accessed December 5, 2005). http://pages.britishlibrary.net/charles.darwin/texts/letters/letters2_02.html (accessed September 14, 2006).

Darwin, F., and A. C. Seward, eds. *More Letters of Charles Darwin.* 2 vols. London, Murray, 1903. Vol. 1, chap. 2, p. 94. http://pages.britishlibrary.net/charles.darwin/texts/more_letters/mletters1_02.html (accessed September 14, 2006).

Davis, P., and D. H. Kenyon. *Of Pandas and People.* Dallas, TX: Haughton, 1993.

Dawkins, R. *The Blind Watchmaker: Why the Evidence of Evolution Reveals a Universe without Design.* New York: Norton, 1986.

———. *Climbing Mount Improbable.* New York: Norton, 1996.

———. "Human Chauvinism." *Evolution* 51, no. 3 (1997): 1015–20.

———. *The Selfish Gene*. New York: Oxford University Press, 1976.

Dembski, W. A. *No Free Lunch: Why Specified Complexity Cannot Be Purchased without Intelligence*. New York: Rowman & Littlefield, 2002.

———. "What Every Theologian Should Know about Creation, Evolution, and Design." *Center for Interdisciplinary Studies Transactions* 3, no. 2 (1995): 1–8.

Demoor, M. "His Way is thro' Chaos and the Bottomless and Pathless: The Gender of Madness in Alfred Tennyson's Poetry." *Neophilologus* 86, no. 2 (2002): 325–35.

D'Errico, F. "Palaeolithic Origins of Artificial Memory Systems." In *Cognition and Material Culture: The Archaeology of Symbolic Storage*, edited by C. Renfrew and C. Scarre, 19–50. Cambridge, UK: MacDonald Institute for Archaeological Research, 1998.

Dobzhansky, T. "Nothing in Biology Makes Sense Except in the Light of Evolution." *American Biology Teacher* 35 (1973): 125–29.

Dunbar, R. I. M. "Neocortex Size as a Constraint on Group Size in Primates." *Journal of Human Evolution* 20 (1992): 469–93.

Durant, W. *The Story of Philosophy*. New York: Simon & Schuster, 1926.

Dykhuizen, D. E. "Santa Rosalia Revisited: Why Are There So Many Species of Bacteria?" *Antonie Van Leeuwenhoek International Journal of General and Molecular Microbiology* 73 (1998): 25–33.

Eldredge, N. "Cretacious Meteor Showers, the Human Ecological 'Niche,' and the Sixth Extinction." In *Extinctions in Near Time: Causes Contexts and Consequences*, edited by R. D. E. MacPhee, 1–14. New York: Kluwer Academic/Plenum Publishers, 1999.

Endler, J. *Natural Selection in the Wild*. Princeton, NJ: Princeton University Press, 1986.

Feldhamer G., J. Whittaker, A. Monty, and C. Weickert. "Charismatic Mammalian Megafauna: Public Empathy and Marketing Strategy." *Journal of Popular Culture* 36, no. 1 (2002): 160–67.

Fettner, A. G. *Viruses: Agents of Change*. New York: McGraw-Hill, 1990.

Forrest, B., and P. R. Gross. *Creationism's Trojan Horse: The Wedge of Intelligent Design*. Oxford: Oxford University Press, 2004.

Frair, W., and P. Davis. *A Case for Creation*. 3rd ed. Chicago: Moody, 1983.

Frank, P. *Einstein, His Life and Times*. New York: Knopf, 1947.

Futuyma, D. J. "Miracles and Molecules." *Boston Review*, February/March 1997. http://www.bostonreview.net/br22.1/futuyma.html (accessed November 20, 2005).

Gage, J. D., and P. A. Tyler. *Deep-Sea Biology: A Natural History of Organisms at the Deep-Sea Floor*. Cambridge: Cambridge University Press, 1996.

Gamble, C. *The Palaeolithic Societies of Europe*. Cambridge: Cambridge University Press, 1999.

Gárate-Lizárraga, I., S. Beltrones, and V. Maldonado-López. "First Record of a *Rhizosolenia debyana* Bloom in the Gulf of California, Mexico." *Pacific Science* 57 no. 2 (2003): 141–45.

Ghiselin, M. T. "The Imaginary Lamarck: A Look at Bogus 'History' in Schoolbooks." *Textbook Letter*, September/October 1994. http://www.textbookleague.org/54marck.htm (accessed June 18, 2005).

Gibbons, A. "Calibrating the Mitochondrial Clock." *Science* 279, no. 5347 (1998): 28–29.

Gould, S. J. *Full House: The Spread of Excellence from Plato to Darwin*. New York: Harmony Books, 1996.

———. *Time's Arrow, Time's Cycle*. Cambridge, MA: Harvard University Press, 1987.

Graves, J. A. M., and M. Westerman. "Marsupial Genetics and Genomics." *Trends in Genetics* 18, no. 10 (2002): 517–21.

Grine, F. E., ed. *Evolutionary History of the Robust Australopithecines*. New York: De Gruyter, 1988.

Harcourt-Smith, W. E. H., and L. C. Aiello. "Fossils, Feet, and the Evolution of Human Bipedal Locomotion." *Journal of Anatomy* 204, no. 5 (2004): 403–16.

Hawkins, M. *Social Darwinism in European and American Thought, 1860–1945: Nature as Model and Nature as Threat*. Cambridge: Cambridge University Press, 1997.

Heesy, C. P. "On the Relationship between Orbit Orientation and Binocular Visual Field Overlap in Mammals." In *Evolution of the Special Senses in Primates*, edited by T. D. Smith, C. F. Ross, N. J. Dominy, and J. T. Laitman. *The Anatomical Record Part A: Discoveries in Molecular, Cellular, and Evolutionary Biology* 281A, no. 1 (2004): 1104–10.

Hildebrand, A. R., G. Penfield, D. Kring, M. Pilkington, A. Zanoguera, S. Jacobsen, and W. Boynton. "Chicxulub Crater: A Possible Cretaceous/

Tertiary Boundary Impact Crater on the Yucatan Peninsula, Mexico." *Geology* 19, no. 9 (1991): 867–71.

Ho, S. Y. W., M. Phillips, A. Cooper, and A. Drummond. "Time Dependency of Molecular Rate Estimates and Systematic Overestimation of Recent Divergence Times." *Molecular Biology and Evolution* 22 (2005): 1561–68.

Hofstadter, R. *Social Darwinism in American Thought*. Boston: Beacon, 1983.

Houle, A. "The Origin of Platyrrhines: An Evaluation of the Antarctic Scenario and the Floating Island Model." *American Journal of Physical Anthropology* 109, no. 4 (1999): 541–59.

Houston, S. D. *The First Writing: Script Invention as History and Process*. Cambridge: Cambridge University Press, 2004.

"Humans and Other Catastrophes." *American Museum of Natural History*. http://www.amnh.org/science/biodiversity/extinction/ (accessed October 8, 2005).

Huxley, T. H. *Evolution and Ethics and Other Essays*. New York: Greenwood, 1968.

Ignatius of Loyola, St. *The Spiritual Exercises of St. Ignatius of Loyola*. Translated from the autograph by Father Elder Mullan, S. J. New York: Kennedy & Sons, 1914. http://www.ccel.org/ccel/ignatius/exercises.pdf (accessed January 10, 2006).

"Intelligent Design Not Science, Says Vatican Newspaper Article." *Catholic News Service*, January 17, 2006. http://www.catholicnews.com/data/stories/cns/0600273.htm (accessed January 18, 2006).

"Invasives Alert! *Anoplophlora glabripennis* (Motchulsky): (Asian Longhorned Beetle)." *Nature Conservancy*. http://tncweeds.ucdavis.edu/alert/alrtanop.html (accessed November 11, 2005).

Irvine, W. *Apes, Angels, and Victorians*. New York: McGraw-Hill, 1955.

Jacobs, B. F. "Palaeobotanical Studies from Tropical Africa: Relevance to the Evolution of Forest, Woodland, and Savannah Biomes." *Philosophical Transactions of the Royal Society of London B* 359 (2004): 1573–83.

Jansen, T., P. Forster, M. Levine, H. Oelke, M. Hurles, C. Renfrew, J. Weber, and K. Olek. "Mitochondrial DNA and the Origins of the Domestic Horse." *Proceedings of the National Academy of Sciences USA* 99, no. 16 (2002): 10905–10.

Janson, H. W. *Apes and Ape Lore in the Middle Ages and the Renaissance*. London: Warburg Institute, 1952.

Ji, Q., Z. Luo, C. Yuan, J. Wible, J. Zhang, and J. Georgi. "The Earliest Known Eutherian Mammal." *Nature* 416 (2002): 816–22.

Johanson, D., and M. A. Edey. *Lucy: The Beginnings of Humankind.* New York: Simon & Schuster, 1981.

Johanson, D., and M. Taieb. "Plio-Pleistocene Hominid Discoveries in Hadar, Ethiopia." *Nature* 260 (1976): 293–97.

Johnson, P. E. *The Wedge of Truth: Splitting the Foundations of Naturalism.* Downers Grove, IL: InterVarsity, 2000.

Jurmain, N., H. Nelson, L. Kilgore, and W. Trevathan. *Introduction to Physical Anthropology*. Belmont, CA: Wadsworth, 1999.

Kaiho, K., Y. Kajiwara, T. Nakano, Y. Miura, H. Kawahata, K. Tazaki, M. Ueshima, Z. Chen, and G. Shi. "End-Permian Catastrophe by a Bolide Impact: Evidence of a Gigantic Release of Sulfur from the Mantle." *Geology* 29, no. 9 (2001): 815–18.

Kavanaugh, M. *A Complete Guide to Monkeys, Apes, and Other Primates*. New York: Viking, 1984.

Kidwell, S. M., and K. W. Flessa. "The Quality of the Fossil Record: Populations, Species and Communities." *Annual Review of Earth and Planetary Sciences* 24 (1996): 433–64.

Kious, W. J., and R. I. Tilling. *The Dynamic Earth: The Story of Plate Tectonics*. Washington, DC: US Government Printing Office, 1996. http://pubs.usgs.gov/publications/text/dynamic.pdf (accessed June 22, 2005).

Klein, R. G. *The Human Career*. Chicago: University of Chicago Press, 1999.

Knoll, A. H. "The Early Evolution of Eukaryotes: A Geological Perspective." *Science* 256, no. 5057 (1992): 622–27.

Kopp, M. E. "Eugenic Sterilization Laws in Europe." *American Journal of Obstetrics and Gynecology* 34 (September 1937): 499–504.

Kropotkin, P. *Mutual Aid: A Factor of Evolution*. Boston: Extending Horizons Books, 1955.

Kunkel, T. A., and K. Bebenek. "DNA Replication Fidelity." *Annual Review of Biochemistry* 69 (2000): 497–529.

Lamarck, J. B. *Zoological Philosophy*. Translated by H. Elliot. Chicago: University of Chicago Press, 1984.

Lambert, D. M., P. Ritchie, C. Millar, B. Holland, A. Drummond, C. Baroni. "Rates of Evolution in Ancient DNA from Adélie Penguins." *Science* 295, no. 5563 (2002): 2270–73.

Landau, M. *Narratives of Human Evolution*. New Haven, CT: Yale University Press, 1991.

Lankester, E. R. *Diversions of a Naturalist*. New York: Macmillan, 1915.

Leakey, R., and R. Lewin. *People of the Lake*. New York: Anchor, 1978.

———. *The Sixth Extinction: Biodiversity and Its Survival*. London: Weidenfeld & Nicolson, 1996.

Lederburg, J., and E. L. Tatum. "Gene Recombination in E. coli." *Nature* 158 (1946): 558.

Lee, A. K., and A. Cockburn. *Evolutionary Ecology of Marsupials*. Cambridge: Cambridge University Press, 1985.

Levine, M. A. "Botai and the Origins of Horse Domestication." *Journal of Anthropological Archaeology* 18, no. 1 (1999): 29–78.

Levy, N. *What Makes Us Moral? Crossing the Boundaries of Biology*. Oxford: Oneworld, 2004.

Lewis, C. S. *The Problem of Pain*. San Francisco: Harper, 2001.

Lodge, D. M. "Biological Invasions: Lessons for Ecology." *Trends in Ecology and Evolution* 8, no. 4 (1993): 133–37.

Lovejoy, A. O. *The Great Chain of Being: A Study of the History of an Idea*. Cambridge, MA: Harvard University Press, 1936.

Lovelock, J. E. *Gaia: A New Look at Life on Earth*. Oxford: Oxford University Press, 1979.

Lucas, J. R. "Wilberforce and Huxley: A Legendary Encounter." *Historical Journal* 22, no. 2 (1979): 313–30.

Lucretius. *On the Nature of the Universe*. Translated and edited by R. E. Latham. London: Penguin Books, 1982.

Lyell, C. *The Antiquity of Man*. London: Dent, 1927.

———. *Principles of Geology: A Facsimile of the First Edition*. Vol. 1. Chicago: University of Chicago Press, 1990.

MacPhee, R. D. E., and C. Flemming. "Requiem Aeternum: The Last Five Hundred Years of Mammalian Species Extinctions." In *Extinctions in Near Time: Causes Contexts and Consequences*, edited by R. D. E. MacPhee, 333–72. New York: Kluwer Academic/Plenum, 1999.

Mallet, J. "A Species Definition for the Modern Synthesis." *Trends in Ecology and Evolution* 10 (1995): 294–99.

Marshall, L. G. "Land Mammals and the Great American Interchange." *American Scientist* 76 (1988): 380–88.

Matsumura, M. "Tennessee Upset: 'Monkey Bill' Law Defeated." *NCSE Reports* 15, no. 4 (1995): 6–7.

May, R. M. "How Many Species?" *Philosophical Transactions of the Royal Society of London B* 330 (1990): 293–304.

Mayr, E. *One Long Argument: Charles Darwin and the Genesis of Modern Evolutionary Thought*. Cambridge, MA: Harvard University Press, 1991.

———. *This Is Biology: The Science of the Living World*. Cambridge, MA: Belknap Press of Harvard University Press, 1997.

———. *What Evolution Is*. New York: Basic Books, 2001.

———. "What Is a Species and What Is Not?" *Philosophy of Science* 63 (1996): 262–77.

Mayr, E., and W. B. Provine, eds. *The Evolutionary Synthesis*. Cambridge, MA: Harvard University Press, 1980.

McHenry, H. M., and C. Coffing. "*Australopithecus* to *Homo*: Transformations in Body and Mind." *Annual Review of Anthropology* 29 (2000): 125–46.

McMullin, E. R., D. C. Bergquist, and C. R. Fisher. "Metazoans in Extreme Environments: Adaptations of Hydrothermal Vent and Hydrocarbon Seep Fauna." *Gravitational and Space Biology Bulletin* 13, no. 2 (2000): 13–24.

Menotti-Raymond, M., and S. J. O'Brien. "Dating the Genetic Bottleneck of the African Cheetah." *Proceedings of the National Academy of Sciences USA* 90, no. 8 (1993): 3172–76.

Miller, K. "Response to Newman." *Perspectives on Science and Christian Faith* 48 (March 1996): 66–68.

Miller, K. R. Review of *Darwin's Black Box: The Biochemical Challenge to Evolution*, by M. J. Behe. *Creation/Evolution* 16, no. 2 (1996): 36–40.

Miller, K. R., and J. Levine, *Biology*. 3rd ed. New Jersey: Prentice-Hall, 1995.

Milne, L . J. *The Balance of Nature*. New York: Knopf, 1960.

Mithen, S. *The Prehistory of the Mind: A Search for the Cognitive Origins of Art, Religion and Science*. London: Thames & Hudson, 1996.

Moorbath, S. "Palaeobiology: Dating Earliest Life." *Nature* 434 (2005): 155.

Moore, G. E. *Principia Ethica*. Cambridge: Cambridge University Press, 1903.

Morris, H. M., and J. C. Whitcomb. *The Genesis Flood*. Philadelphia: Presbyterian and Reformed Publishing, 1961.

Murphy, W. J., E. Eizirik, W. Johnson, Y. Zhang, O. Ryder, and S. J. O'Brien. "Molecular Phylogenetics and the Origins of Placental Mammals." *Nature* 409 (2001): 614–18.

Nikaido, M., A. P. Rooney, and N. Okada. "Phylogenetic Relationships among Cetartiodactyls Based on Insertions of Short and Long Interpersed Elements: Hippopotamuses Are the Closest Extant Relatives of Whales." *Proceedings of the National Academy of Sciences USA* 96 (1999): 10261–66.

O'Brien, S.J., M. Roelke, L. Marker, A. Newman, C. Winkler, D. Meltzer, L. Colly, J. Evermann, M. Bush, and D. Wildt. "Genetic Basis for Species Vulnerability in the Cheetah." *Science* 227, no. 4693 (1985): 1428–34.

Officer, C., and G. Page. *The Great Dinosaur Extinction Controversy*. New York: Addison-Wesley, 1996.

Orr, H. A. "Darwin v. Intelligent Design (Again): The Latest Attack on Evolution Is Cleverly Argued, Biologically Informed—and Wrong." *Boston Review* (December 1996/January 1997): 28–31. http://www.bostonreview.net/BR21.6/orr .html (accessed November 20, 2005).

Ota, R., and D. Penny. "Estimating Changes in Mutational Mechanisms of Evolution." *Journal of Molecular Evolution* 57 (2003): S233–S240.

*Oxford English Dictionary*. 2nd ed. Vol. 9. Oxford: Oxford University Press, 1989.

Oxnard, C. *Fossils, Teeth and Sex: New Perspectives on Human Evolution*. Seattle: Washington University Press, 1987.

Pagel, M., and W. Bodmer. "A Naked Ape Would Have Fewer Parasites." *Biology Letters* 270 (2003): S117–S119.

Paley, W. *Natural Theology*. New York: American Tract Society, n.d.

Pennock, R. T., ed. *Intelligent Design Creationism and Its Critics: Philosophical, Theological, and Scientific Perspectives*. Cambridge, MA: MIT Press, 2001.

———. "The Prospects for a 'Theistic Science.'" *Perspectives on Science and Christian Faith* 50 (September 1998): 205–209.

———. *Tower of Babel: The Evidence against the New Creationism*. Cambridge, MA: Bradford Books/MIT Press, 1999.

———. "The Wizard of ID: Reply to Dembski." In *Intelligent Design Creationism and Its Critics: Philosophical, Theological, and Scientific Perspectives*, edited by R. T. Pennock, 645–67. Cambridge, MA: MIT Press, 2001.

Pierce, S. K., T. Maugel, M. Rumpho, J. Hanten, and W. Mondy. "Annual Viral Expression in a Sea Slug Population: Life Cycle Control and Symbiotic Chloroplast Maintenance." *Biological Bulletin* 197, no. 1 (1999): 1–6.

Pimm, S. *The Balance of Nature? Ecological Issues in the Conservation of Species and Communities*. Chicago: University of Chicago Press, 1991.

Pine, R. H. "New Mammals Not So Seldom." *Nature* 368 (1994): 593.

Prigogene, I., and J. Stengers. *Order Out of Chaos: Man's New Dialogue with Nature*. New York: Bantam Books, 1984.

Rachels, J. *The Elements of Moral Philosophy*. 2nd ed. New York: McGraw-Hill, 1993.

Raup, D. *Extinction: Bad Genes or Bad Luck?* New York: Norton, 1991.

Raup, D., and J. Sepkoski. "Periodicity of Extinctions in the Geologic Past." *Proceedings of the National Academy of Sciences USA* 81 (1984): 801–805.

Ray, J. *The Wisdom of God Manifested in the Works of the Creation: In Two Parts*. London: Innys, 1717.

Richardson, D. M., P. Pyšek, M. Rejmánek, M. Barbour, F. Panetta, and C. West. "Naturalization and Invasion of Alien Plants: Concepts and Definitions." *Diversity and Distributions* 6 (2000): 93–107.

Rothwell, T. "Evidence for Taming of Cats." *Science* 305, no. 5691 (2004): 1714.

Rupke, N. A. *The Great Chain of History: William Buckland and the English School of Geology (1814–1849)*. New York: Oxford University Press, 1983.

Sagan, C. *A Demon-Haunted World: Science as a Candle in the Dark*. New York: Ballantine Books, 1996.

Sagan, C., and A. Druyan. *Shadows of Forgotten Ancestors: A Search for Who We Are*. New York: Ballantine Books, 1992.

Schluter, D. "Ecological Causes of Adaptive Radiation." *American Naturalist* 148 (November 1996): S40–S64.

Schopf, J. W., and B. M. Packer. "Early Archean (3.3 Billion to 3.5 Billion-Year-Old) Microfossils from Warrawoona Group, Australia." *Science* 237 (1987): 70–73.

"Science and Technology: Public Attitudes and Understanding." In *Science and Engineering Indicators*, chapter 7. Arlington, VA: National Science Foundation, Division of Science Resources and Statistics, 2004. http://www.nsf.gov/statistics/seind04/pdf/c07.pdf (accessed May 5, 2005).

Scott, E. C. *Evolution vs. Creationism*. Westport, CT: Greenwood, 2004.

———. "State of Alabama Distorts Science, Evolution." *NCSE Reports* 15, no. 4 (1995): 10–11.

Serpell, J., and P. Barrett, eds. *The Domestic Dog: Its Evolution, Behaviour, and Interactions with People*. Cambridge: Cambridge University Press, 1995.

Shadewald, R. J. "The Evolution of Bible-Science." In *Scientists Confront Creationism*, edited by L. R. Godfrey, 283–99. New York: Norton, 1983.

Sheehan, P. M., D. Fastovsky, C. Barreto, and R. Hoffmann. "Dinosaur Abundance Was Not Declining in a '3 m gap' at the Top of the Hell Creek Formation, Montana and North Dakota." *Geology* 28, no. 6 (2000): 523–26.

Shermer, M. *The Science of Good and Evil: Why People Cheat, Gossip, Care, Share, and Follow the Golden Rule*. New York: Holt, 2004.

Shirer, W. L. *The Rise and Fall of the Third Reich: A History of Nazi Germany*. New York: Simon & Schuster, 1960.

Simpson, G. G. "How Many Species?" *Evolution* 6, no. 3 (1954): 342.

Smith, J. M., and E. Szathmáry. *The Major Transitions in Evolution*. Oxford: Freeman, 1995.

Solar Physics Group (NASA). "Sun Facts." http://science.msfc.nasa.gov/ssl/pad/solar/sunspots.htm (accessed August 24, 2005).

Spencer, H. *Social Statics: The Conditions Essential to Happiness Specified, and the First of Them Developed*. New York: Robert Schalkenbach Foundation, 1970.

Stapleton, A. E. "Ultraviolet Radiation and Plants: Burning Questions." *Plant Cell* 4 (1992): 1353–58.

Steiper, M. E., N. M. Young, and T. Y. Sukarna. "Genomic Data Support the Hominoid Slowdown and an Early Oligocene Estimate for the Hominoid Cercopithecoid Divergence." *Proceedings of the National Academy of Sciences USA* 101, no. 49 (2004): 17021–26.

Stephensen, S. L., and H. Stempen. *Myxomycetes: A Handbook of Slime Molds*. Portland, OR: Timber, 2000.

Sterelny, K. *Dawkins vs. Gould: Survival of the Fittest*. Cambridge, UK: Icon Books, 2001.

Strickberger, M. W. *Genetics*. 3rd ed. New York: Macmillan, 1985.

Sullivan, C., and C. M. Smith. "Getting the Monkey off Darwin's Back: Four Common Myths about Evolution." *Skeptical Inquirer* (May/June 2005): 43–48.

Suzuki, D. *The Sacred Balance: Rediscovering Our Place in Nature*. Toronto: Greystone Books, 2002.

Takai, M., F. Anaya, N. Shigehara, and T. Steoguchi. "New Fossil Materials of the Earliest New World Monkey, *Branisella boliviana*, and the Problem of Platyrrhine Origins." *American Journal of Physical Anthropology* 111, no. 2 (2000): 263–81.

Thaxton, C. B., W. L. Bradley, and R. L. Olson. *The Mystery of Life's Origin: Reassessing Current Theories*. New York: Philosophical Library, 1984.

Thomas, C. D. "Fewer Species." *Nature* 347 (1990): 237.

Thompson, B. H. "Where Have All My Pumpkins Gone? The Vulnerability of Insect Pollinators." In *Food and Agricultural Security: Guarding Against Natural Threats and Terrorist Attacks Affecting Health, National Food Supplies, and Agricultural Economics*, edited by T. W. Frazier and D. C. Richardson. *Annals of the New York Academy of Sciences* 894 (1999): 189–98.

Trinkaus, E., and M. Zimmerman. "Trauma Among the Shanidar Neanderthals." *American Journal of Physical Anthropology* 57, no. 1 (1982): 61–76.

Turney, J. *Lovelock and Gaia: Signs of Life*. New York: Columbia University Press, 2003.

Tyndall, J. *Address Delivered before the British Association Assembled at Belfast, with Additions*. London: Longmans & Green, 1874.

Ursing, B. M., and U. Arnason. "Analyses of Mitochondrial Genomes Strongly Support a Hippopotamus-Whale Clade." *Proceedings of the Royal Society of London B* 265 (1998): 2251–55.

Ussery, D. "Darwin's Transparent Box: The Biochemical Evidence for Evolution." In *Why Intelligent Design Fails: A Scientific Critique of the New Creationism*, edited by M. Young and T. Edis, 48–57. New Brunswick, NJ: Rutgers University Press, 2005.

Van Valen, L. "A New Evolutionary Law." *Evolutionary Theory* 1 (1973): 1–30.

Waal, F. B. M. de. *Good Natured: The Origins of Right and Wrong in Humans and Other Animals*. Cambridge, MA: Harvard University Press, 1996.

Ward, B. B. "How Many Species of Prokaryotes Are There?" *Proceedings of the National Academy of Sciences USA* 99, no. 16 (1999): 10234–36.

Wenke, R. J. *Patterns in Prehistory: Humankind's First Three Million Years*. New York: Oxford University Press, 1999.

Wharton, D. A. *Life at the Limits: Organisms in Extreme Environments*. Cambridge: Cambridge University Press, 2002.

"What's Wrong with 'Theory Not Fact' Resolutions." *National Center for Science Education* (December 7, 2000). http://www.ncseweb.org/resources/articles/8643_whats_wrong_with_theory_not_12_7_2000.asp (accessed June 12, 2005).

Wheeler, P. E. "The Evolution of Bipedality and Loss of Functional Body Hair in Humans." *Journal of Human Evolution* 13 (1984): 91–98.

Wildman, D. E., M. Uddin, G. Liu, L. I. Grossman, and M. Goodman. "Implications of Natural Selection in Shaping 99.4% Nonsynonymous DNA Identity between Humans and Chimpanzees: Enlarging Genus *Homo*." *Proceedings of the National Academy of Sciences USA* 100 (2003): 7181–88.

Williamson, M. *Biological Invasions*. Population and Community Series 15. London: Chapman & Hall, 1996.

Wilson, C. *The Invisible World: Early Modern Philosophy and the Invention of the Microscope*. Princeton, NJ: Princeton University Press, 1995.

Wilson, E. O. *Sociobiology*. Cambridge, MA: Harvard University Press, 1977.

Wilson, R. A., ed. *Species: New Interdisciplinary Essays*. Cambridge, MA: MIT Press, 1999.

Wood, B., and M. Collard. "Is *Homo* Defined by Culture?" *Proceedings of the British Academy of Sciences* 99 (1999): 11–23.

World Health Organization. *Ebola Haemorrhagic Fever Fact Sheet*. http://www.who.int/mediacentre/factsheets/fs103/en/ (accessed December 12, 2005).

Wright, S. *Evolution and the Genetics of Populations: A Treatise*. Chicago: University of Chicago Press, 1968.

Young, M., and T. Edis, eds. *Why Intelligent Design Fails: A Scientific Critique of the New Creationism*. New Brunswick, NJ: Rutgers University Press, 2005.

# Index